Eastern Snow Conference 2017

AA004428

Ottawa, Canada
6 - 8 June 2017

ISBN: 978-1-7138-0356-0

Printed from e-media with permission by:

Curran Associates, Inc.
57 Morehouse Lane
Red Hook, NY 12571

CURRAN ASSOCIATES INC. proceedings.com

Some format issues inherent in the e-media version may also appear in this print version.

Copyright© (2017) by Eastern Snow Conference
All rights reserved.

Printed with permission by Curran Associates, Inc. (2020)

For permission requests, please contact Eastern Snow Conference
at the address below.

Eastern Snow Conference
C/O Dr. Krystopher Chutko
117 Science Pl-Dept. Geography
Saskatoon, Sk, Canada S7N 5C8

https://www.easternsnow.org/

Additional copies of this publication are available from:

Curran Associates, Inc.
57 Morehouse Lane
Red Hook, NY 12571 USA
Phone: 845-758-0400
Fax: 845-758-2633
Email: curran@proceedings.com
Web: www.proceedings.com

Proceedings of the
74th ANNUAL

EASTERN SNOW CONFERENCE

6-8 June 2017

University of Ottawa, Ottawa
Ontario, Canada

CONTENTS

FOREWORD

This proceedings volume contains papers presented at the 74th Eastern Snow Conference (ESC) held 6-8 June 2017 in Ottawa, Ontario, Canada. The meeting featured oral and poster sessions on recent advances in remote sensing, remote sensing theory, applications of remote sensing ranging from the ice sheets to the mid-latitudes, snow and ice processes, and hydroclimatology.

The ESC is a joint United States and Canadian forum for discussing recent work on operational, applied and scientific issues related to snow and ice. It also retains an increasing interest as a symposium where novel approaches to cryospheric science of international significance are presented. The ESC has published an annual series of proceedings since 1952. Typical topics include studies of snow and ice as materials, snow removal, meteorological forecasting, river ice control, snow hydrology, snow chemistry, glaciology, remote sensing of snow and ice, and snow ecology. Membership in the ESC is open to all interested individuals and corporations. Additional copies of the current proceedings and all back issues can be obtained from the Secretary. Additional information about the Eastern Snow Conference may be found at: http://www.easternsnow.org/.

We thank members of the ESC executive committee (especially Mauri Pelto) for outstanding effort in publishing ESC 2016 proceedings in *Physical Geography*. Over the years, the proceedings committee has contributed considerable time to this publication and has enhanced the quality of these proceedings.

The 2017 meeting of the Eastern Snow Conference and these Proceedings were made possible in part by Corporate membership from Campbell Scientific Canada and Geonor. The Byrd Polar & Climate Research Center at The Ohio State University provided meeting space, technical, and logistic support.

Campbell Scientific (Canada) Corp.
Edmonton, Alberta, Canada
http://www.campbellsci.ca

Hoskin Scientific Limited
Burnaby, British Columbia, Canada
http://www.hoskin.ca

GEONOR
Augusta, New Jersey, USA
http://www.geonor.com

Laboratory for Cryospheric Research
Ottawa, Ontario, Canada
https://cryospheric.org/

University of Ottawa
Ottawa, Ontario, Canada
https://www.uottawa.ca/en

We look forward to seeing many of you at the 75th Eastern Snow Conference meeting to be held 5–8 June 2018 at the NOAA Center for Weather and Climate Prediction in College Park, Maryland, USA. To commemorate the 75th meeting, the scientific program will be expanded to celebrate all things snow. The general theme of the meeting will be "Snow: Past, Present, and Future".

Alexandre Langlois and Krystopher Chutko
ESC Proceedings Co-Editors
Université de Sherbrooke and University of Saskatchewan

STATEMENT OF PURPOSE

The Eastern Snow Conference (ESC) is a joint Canadian/US organization founded in the 1940s, originally with members primarily from eastern North America. Our current members are scientists, snow surveyors, engineers, technicians, professors, students, and operational and maintenance professionals from North America, the United Kingdom, Japan, and Germany. There is a western counterpart to the ESC, the Western Snow Conference (WSC), which also is a joint Canadian/US organization. Every fifth year, the ESC and the WSC hold joint meetings.

The Eastern Snow Conference is a forum that brings the research and operations communities together to discuss recent work on scientific, applied, and operational issues related to snow and ice. The location of the conference alternates yearly between the United States and Canada, and attendees present their work by either giving a talk or presenting a poster. Most resulting papers are reviewed, edited, and published in our yearly *Proceedings of the Eastern Snow Conference*. In recent years, the ESC meetings have included sessions on snow physics, winter survival of animals, snow and ice loads on structures, river ice, remote sensing of snow and ice, and glacier processes. Volumes of the *Proceedings* can be found in libraries throughout North America and Europe, and the papers are also available through the National Technical Information Service (NTIS) in the United States and CISTI in Canada.

* * * * * * * * *

Le Colloque sur la neige région Est (ESC) est une organisation américaine-canadienne fondée dans les années '40 et dont les membres provenaient à l'origine principalement de l'est de l'Amérique-du-Nord. Actuellement, les membres, qu'ils soient chercheurs, techniciens, ingénieurs, professeurs, étudiants, et spécialistes des services d'exploitation et d'entretien, viennent non seulement d'Amérique-du-Nord, mais aussi du Royaume-Uni, du Japon, et d'Allemagne. Le Colloque sur la neige-région Ouest (WSC), aussi une organisation américaine-canadienne, est l'homologue de l'ESC pour l'ouest nord-américain. Tous les cinq ans, l'ESC et la WSC organisent des réunions en commun.

L'ESC est un forum qui rassemble chercheurs et responsables des services d'exploitation pour discuter des travaux récents sur les problèmes scientifiques, opérationnels, ou autres dus à la neige et à la glace. Le site de cette réunion annuelle alterne entre les États Unis et le Canada. Les participants y présentent les résultats de leurs travaux par des communications orales ou au moyen d'affiches. Ces communications, une fois revues et éditées, sont publiées dans les *Annales de l'ESC*. Dans les années récentes, les réunions de l'ESC ont inclus des sessions sur la physique de la neige, la survie hivernale de la faune, les forces exercées par la neige et la glace sur les structures et les bâtiments, la glace de rivière, la télédétection de la neige et de la glace, et les processus glaciaires. Les *Annales de l'ESC* sont accessibles dans la plupart des bibliothèques scientifiques d'Amérique-du-Nord et d'Europe. Des copies d'articles peuvent être obtenues du National Technical Information Service (NTIS) aux États Unis et son équivalent au Canada, le CISTI.

EXECUTIVES FOR THE 2016-17 74th EASTERN SNOW CONFERENCE

PRESIDENT

Michael Durand
Columbus
Ohio, USA

PAST PRESIDENT

Alain Royer
Sherbrooke
Quebec, Canada

VICE-PRESIDENT AND PROGRAM CHAIR

Alexandre Roy
Montreal
Quebec, Canada

SECRETARY-TREASURER (Canada)

Miles Ecclestone
Peterborough
Ontario, Canada

Krystopher Chutko
Saskatoon
Saskatchewan, Canada

Alexandre Langlois
Sherbrooke
Quebec, Canada

SECRETARY-TREASURER (USA)

Ken Rancourt
Conway
New Hampshire, USA

Derrill Cowing
Monmouth
Maine, USA

ESC PROCEEDINGS EDITORS

Alexandre Langlois
Sherbrooke
Quebec, Canada

Krystopher Chutko
Saskatoon
Saskatchewan, Canada

***PHYSICAL GEOGRAPHY* EDITORS**

Mauri Pelto (Chair)
Dudley
Massachusetts, USA

Robert Hellström
Bridgewater
Massachusetts, USA

STEERING COMMITTEE

George Riggs (Chair)
Gambrills
Maryland, USA

Chris Fuhrman
Chapel Hill
North Carolina, USA

Laura Thomson
Ottawa
Ontario, Canada

Steve Howell
Toronto
Ontario, Canada

Craig Smith
Saskatoon
Saskatchewan, Canada

Eli Deeb
Hanover
New Hampshire, USA

Carrie Vuyovich
Hanover
New Hampshire, USA

Weisnet Medal for Best Student Paper

Meteorological inventory of Rain-On-Snow events and detection assessment in the Canadian Arctic Archipelago using passive microwave radiometry

DOLANT C., [1,2] LANGLOIS A. [1,2], BRUCKER L.[3,4], ROYER A. [1,2], ROY A. [1,2], AND MONTPETIT B. [1]

ABSTRACT

The Arctic has witnessed significant warming over the past four decades, leading to a variability of consequences such as heat waves, increased occurrence of winter storms and rain-on-snow events (ROS). The spatial and temporal distributions of ROS across the Canadian Arctic Archipelago (CAA) remain poorly understood owing to their sporadic nature in time and space, which motivated the development of remote sensing algorithms. The potential of passive microwave observations was demonstrated in detecting ROS events. Lately, promising results were obtained using an algorithm based a brightness temperature gradient ratio. Its validation however remained limited due to a short study period and limited number of sites. This paper uses a large meteorological dataset across the CAA to further adapt our existing algorithm. Hence, this study highlights the distribution and evolution of ROS occurrences reported since 1985 at 14 Environment and Climate Change Canada (ECCC) weather stations across the CAA. Data show that more than 600 ROS events were inventoried since 1985, 80% of which occurred during the spring season. We introduce an adaptation of a detection algorithm (by sensitivity analysis on the detection threshold) with an error of ~5%, to investigate spatiotemporal patterns in event occurrence across the CAA.

Keywords: Rain-on-snow events; Canadian Arctic Archipelago (CAA); climatological trends; microwave radiometry

INTRODUCTION

The impacts of global climate change on the environment are significant due to a variety of climate feedback processes, such as an increased occurrence of heat waves, winter storms (Trenberth et al., 2007; Trenberth, 2011) and rain-on-snow (ROS) events (Brown and Mote, 2009; Chen et al. 2013). This is especially true in northern regions, where the warming is most pronounced (Serreze et al. 2009; Liston and Hiemstra, 2011; Serreze and Barry, 2011; Pradhanang et al. 2012 & 2013; Klos et al. 2014; Cullather et al. 2016). Direct consequences in the northern hemisphere include an overall reduction of snow depths (Brown and Braaten, 1998), snow cover extent (Brown et al. 2010; Derksen and Brown, 2012; Derksen et al., 2012) and sea ice extent (Steele et al. 2008; Holland et

[1] Centre d'Application et de Recherches en Télédétection, Université de Sherbrooke, Qc, Canada
[2] Centre for Northern Studies, Qc, Canada
[3] NASA Goddard Space Flight Center, Cryospheric Sciences Laboratory, Greenbelt, MD 20771, USA
[4] Universities Space Research Association, Goddard Earth Sciences Technology and Research Studies and Investigations, Columbia, MD 21044, USA

al. 2010; 2014; Comiso J.C, 2014), and an increased occurrence of extreme winter events (Grassi et al., 2013). More specifically, the rise in temperatures (L'hôte, 2005) has significant impacts on the precipitation regime (phase) and is amongst the most significant consequences of atmospheric warming and variability in the Arctic since the early 1980s (Winton, 2006; Liston and Hiemstra, 2011; IPCC, 2007 & 2014; Langlois et al. 2017). Of note, under warming temperature trends, ROS events on the one hand could increase with increasing ratio of rain versus snowfall events while on the other hand could decrease if the snow cover decreases. Such competitive impacts may render the trend of ROS difficult to analyse (Chen et al., 2015; Ye et al., 2008). However, from ROS-altered snow state analysis using satellite data, ROS events appear now more frequent than during the 1990s (Langlois et al. 2017) ROS significantly alters the snowpack state. Firstly, it leads to an increased liquid water content (LWC) in the snowpack. This process when followed by low air temperatures can then be responsible for the formation of ice crusts (Montpetit et al. 2013; Montpetit, 2015; Dolant et al. 2016) that have a strong impact on ecology (e.g., ungulate grazing conditions) (Putkonen and Roe, 2003; Rennert et al. 2009; Bokhorst et al. 2016; Johnson et al. 2016; Sokolov et al. 2016; Langlois et al. 2017; Ouellet et al. 2017), hydrology (e.g., modification of flow, soil saturation), and energy balance (e.g., modification of snow and soil surface and permafrost) (Putkonen, 1998; Dethloff et al. 2006; Mazurkiewicz et al., 2008; Romanovsky et al. 2010). ROS can also impact the snow surface properties (e.g., albedo) and the properties of the snow layers, such as density (Marshall et al., 1999), liquid water content and thermal conductivity (Domine et al., 2016).

Detecting ROS events from space borne microwave radiometry has been the subject of several studies (Grenfell and Putkonen, 2008; Dolant et al., 2016; Langlois et al., 2017). The use of passive microwaves (PMW) makes it possible to obtain information on the different snowpack layers based on their dielectric response. This represents an interesting avenue for tracking and studying ROS events across the Arctic from space. Grenfell and Putkonen (2008) demonstrated the possibility of using PMW to monitor ROS from space with a combination of microwave brightness temperature (T_B) gradient and polarization ratios. Their study was about on a single ROS event on Banks Island, and the authors highlighted the need for more statistics-based research. This motivated the work by Dolant et al. (2016), who developed an empirical approach to detect ROS using the gradient ratio in both vertical and horizontal polarizations. That work was validated with ROS observations in Nunavik (northern Québec, Canada), but the validation remained spatially and temporally limited. More recently, Langlois et al. (2017) analyzed the occurrence of ROS and ice layers detected from PMW and the link between these occurrences and caribou populations. Their work included PMW signatures from 18 islands across the Canadian Arctic Archipelago (CAA), and suggested an adjustment to the threshold initially proposed in Dolant et al. (2016).

The objectives of this paper are: (1) to inventory and classify all ROS observations recorded by the network of Environment and Climate Change Canada meteorological stations in the CAA and conduct a statistical analysis on ROS occurrence trends; and (2) to use this new dataset of ROS observations to validate and adapt the gradient ratio threshold suggested by Dolant et al. (2016) and evaluate the best value minimizing omission and commission errors in an operational application.

STUDY SITES AND DATA

Meteorological data

Water daily amount

ECCC characterizes the liquid precipitation by raindrop size and velocity to distinguish between precipitation phases. Here, we only present information about phase distinction (i.e., drizzle, hail, snow, rain) and precipitation amount (cumulative of the day) (Fig. 1A). Liquid water particles are discriminated as rain when the raindrop diameter is larger than 0.5 mm. Raindrops are normally larger in size than drizzle particles (i.e., small water drops with a diameter less than 0.5mm, uniform precipitation), however the partial evaporation of rain drops can lead to confusion. Solid precipitation (i.e., snow, hail), which can also blend with liquid precipitation, was classified as ROS $_{mix}$ in this study. Precipitation amounts were measured in millimeters with a rain gauge sensor, and values below 0.2 mm were classified as "Trace". Precipitation was also characterized by intensity, from low (precipitation rate less than or equal to 2.5 $mm.h^{-1}$) to heavy (precipitation rate equal to or greater than 7.5 $mm.h^{-1}$).

In our database, precipitation events had predominantly small water amount (Fig. 1A). In all four classes (ROS_{rain}, $ROS_{rain/drizzle}$, ROS_{mix} and $ROS_{Unclass}$), the median amount of precipitation is under 0.9 mm while the mean amount precipitation ranges between 1.3 mm and 2.1 mm. The lowest average precipitation was in Alert (AT) and Sachs Harbour (SH), with 0.78±0.63 mm and 0.82±1.78 mm of rain per event, respectively. The highest values (>3mm) were recorded in Pond Inlet (PI), Nanisivik (NK) and Kugaaruk (KK), with precipitation averages of 3.6±5.60, 3.7±3.15 and 4.64±4.33 mm of rain, respectively.

Air temperature

The station temperature data (Fig. 1B) includes four distinct measurements: dry thermometer, wet thermometer and daily maxima and minima thermometers. These measurements are accurate to one tenth of a degree Celsius. The air temperatures in Figure 1 were extracted and averaged during ROS events only (ROS_{rain}, $ROS_{rain/drizzle}$, ROS_{mix} and $ROS_{Unclass}$). For our study period, the results showed that ROS_{rain} was observed at temperatures of 1.41±2.14 °C while $ROS_{rain/drizzle}$, ROS_{mix} and $ROS_{Unclass}$, occurred at colder temperatures, 0.06±2.56 °C, -1.06±6.8 °C and -2.11±5.05 °C, respectively. When considering all four classes, the mean observed air temperature during ROS was 0.05±3.77 °C).

Snow depth measurement

Snow depth (Fig. 1C) was measured by a trained person at the station using a snow scale (cm), and hourly measurements were available. The minimum observed snow depth we used was 5 cm to limit the number of satellite observations over grid cells with partial snow cover. For our study period, the statistics for snow depth measurements are similar for the four classes: 17.7±14.4 mm, 16.1±12.1 mm, 20.5±15.0 mm and 17.5±13.9 mm for ROS_{rain}, $ROS_{rain/drizzle}$, ROS_{mix} and $ROS_{Unclass}$ conditions, respectively.

Rain-on-Snow duration

The complementary observations recorded ROS event duration and precipitation type. An automatic system qualified the precipitation by state: rain, snow, drizzle, hail, freezing rain, freezing drizzle and unknown. We also extracted ROS event duration data. Results show ROS duration of 4.8±3.7 h, 6.0±4.3 h, 5.0±3.8 h and 2±0 h for the ROS_{rain}, $ROS_{rain/drizzle}$, ROS_{mix} and $ROS_{Unclass}$ conditions, respectively.

Data on precipitation intensity (low, moderate and heavy) was also available. Other information such as visibility and cloud conditions was available, but was not used in this study.

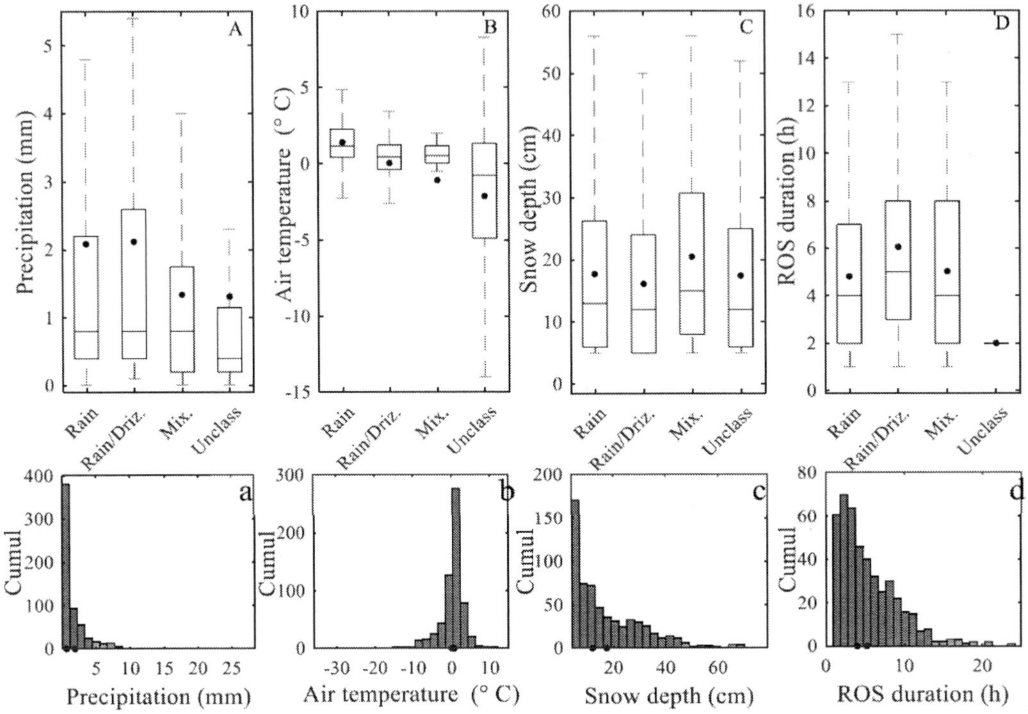

Figure 1. Statistics (upper row, labelled with capital letters) and distribution (lower row, labelled with lowercase letters) of precipitation amount (A,a), air temperature (B,b), snow depth (C,c) and ROS duration (D,d) measured at ECCC meteorological station. The values represent the average of all sites for four classes of ROS events. The boxplot (capital letter) represents interquartile range; black lines represent median and points represent mean value. The histograms (lowercase letters) show a distribution of ROS events (all classes); black points represent median value and blue points represent mean value.

Passive microwave data

For this study, time series of T_B were recorded by three passive microwave sensors onboard five different platforms between August 29, 1984 and June 18, 2014, and were downloaded from the National Snow and Ice Data Center (NSIDC) and used here for the ROS detection retrievals. The T_B dataset at 19 and 37 GHz in both horizontal H and vertical V polarizations was obtained from the Scanning Multichannel Microwave Radiometer (SMMR) (Knowles et al., 2000) onboard the Nimbus-7 platform for the period August 29, 1984 to July 9, 1987. After this period, the T_B was extracted from the Special Sensor Microwave/Imager (SSM/I) onboard Defense Meteorological Satellite Program (DMSP) satellite F08 from July 9, 1987 to December 6, 1991, and the SSM/I and the Special Sensor Microwave Imager/Sounder (SSMIS) (Wentz, 1997; Wentz and Spencer, 1998, Colton and Poe, 1999; Wentz, 2013) onboard DMSP F11, F13 and F17 from December 6, 1991 to June 18, 2014. The spatial resolution of these products is 25 km, projected on the Equal-Area Scalable Earth Grid (EASE-Grid). The products used consist of daily averages in EASE-Grid pixels for ascending and descending orbits $T_{B\ asc}$ and $T_{B\ desc}$). The ascending node is at local noon for SMMR, at 6:00 am for SSM/I F08 and 6:00 pm for T_B extracted from SSM/I and SSMI/S F11, F13 and F17 (local time). The descending node is offset by 12 hours.

METHODS

Inventory approach

Our inventory analysis focuses on the period 1984 (start of winter) to 2014 (end of winter), a period with coincident meteorological measurements (precipitation, air temperature, and snow depth) and human observations (qualitative information about precipitation type and ROS duration) from 14 ECCC stations across the CAA as well as daily passive microwave observations.

For each station and event during the analysis period, we extracted hourly measurement of precipitation amount, air temperature and snow depth values. Furthermore, in this study, four precipitation classes were analyzed: 1) precipitation composed of rain only ($ROS_{rain.}$); 2) precipitation composed of rain and drizzle ($ROS_{rain/drizzle}$); 3) precipitation composed of rain, hail and snow ($ROS_{mix.}$); and 4) rain with no specific observation of phase ($ROS_{Unclass}$). This classification is necessary to evaluate and enhance the detection capacity of our retrieval approach.

Atmospheric correction

An atmospheric correction is needed to remove atmospheric contribution in the satellite signal. Precipitable water data (PWAT) from the North American Regional Reanalysis (NARR) (Mesinger et al., 2006) was used to correct T_B data for atmospheric contributions (Roy et al., 2012). Atmospheric correction was calculated using the Millimeter-wave propagation model (Liebe, 1989) implemented in the Helsinki University of Technology (HUT) snow emission model (Pulliainen et al., 1999). The model considers radiative transfer through the atmospheric layers and calculates values of downwelling T_B ($T_{Batm\downarrow}$), atmospheric transmissivity (γ_{atm}) and atmospheric upwelling T_B ($T_{Batm\uparrow}$) (Liebe, 1989).

The relationship between PWAT and atmospheric upwelling contributions (adjusted from Roy, 2013) were expressed as follows:

$$T_{B_{atm}\uparrow(19)} = (0.879 \cdot PWAT) + 4.780 \quad (1)$$

$$T_{B_{atm}\uparrow(37)} = (0.602 \cdot PWAT) + 18.297$$

(2)

The T_B corrected from the atmospheric contribution was obtained using Eq. (3):

$$T_{Bf\ corrected}^{pol} = T_{Bf\ raw}^{pol} \cdot \tau_f + T_{B_{atm}\uparrow f} \quad (3)$$

with

$$\tau_{(19)} = (-0.003 \cdot PWAT) + 0.979 \quad (4)$$

$$\tau_{(37)} = (-0.002 \cdot PWAT) + 0.924 \quad (5)$$

Over the CAA, the atmospheric corrections led to corrections averaging 1±0.4 K, 2±0.6 K, 1.6 ±0.4 K, 4±0.8 K for T_B at 19 GHz V and H and 37 GHz V and H, respectively.

Rain-On-Snow detection algorithm

The presence of liquid water within the snow and on its surface increases emissivity proportionally to frequency. As such, warm water at the surface (from a ROS event) will lead to a rapid increase of T_B at 37 GHz (faster than 19 GHz). Dolant et al. (2016) demonstrated that in the horizontal polarization, the T_B at 37 GHz becomes warmer than at 19 GHz, while 19 GHz in the vertical polarization remains warmer owing to a higher penetration depth (soil contributions). Dolant et al. (2016) demonstrated that the gradient ratio used by Grenfell and Putkonen (2008) in such conditions is positive in the vertical polarization and negative in the horizontal polarization.

$$GRP = \frac{T_{B37}^{v} - T_{B19}^{v}}{T_{B37}^{v} + T_{B19}^{v}} \cdot \frac{T_{B37}^{h} + T_{B19}^{h}}{T_{B37}^{h} - T_{B19}^{h}} \quad (6)$$

The ratio of gradient ratio, named GRP (Eq. 6), gives a signal inversion (negative GRP value) which characterizes rainfall. Dolant et al. (2016) demonstrated that a GRP threshold at 1 allowed for ROS detection in Nunavik, but it may be slightly too high for a larger scale application in the Arctic as considered in this paper.

RESULTS AND DISCUSSION

Rain-On-Snow analysis from meteorological stations

Using 14 weather stations in the CAA, a total of 625 ROS events were identified, of which 265 were classified as ROS_{rain}, 174 as $ROS_{rain/drizzle}$, 31 as ROS_{mix} and 155 as $ROS_{Unclass}$ for the 1984-2014 period. Figure 2A shows the distribution of ROS classes across the CAA and highlights a higher occurrence of ROS at lower latitudes (< 70°N). High latitude stations (> 70°N) registered less than 150 ROS events between 1984 and 2014 (i.e., ~22% of all ROS events identified). The highest number of ROS occurrences was seen at Hall Beach (HB), with more than 100 events throughout the period. This variability between northern and southern arctic stations could be explained by oceanic currents, warmer seasonal temperatures and associated increased moisture, but no data are available to validate such assumption.

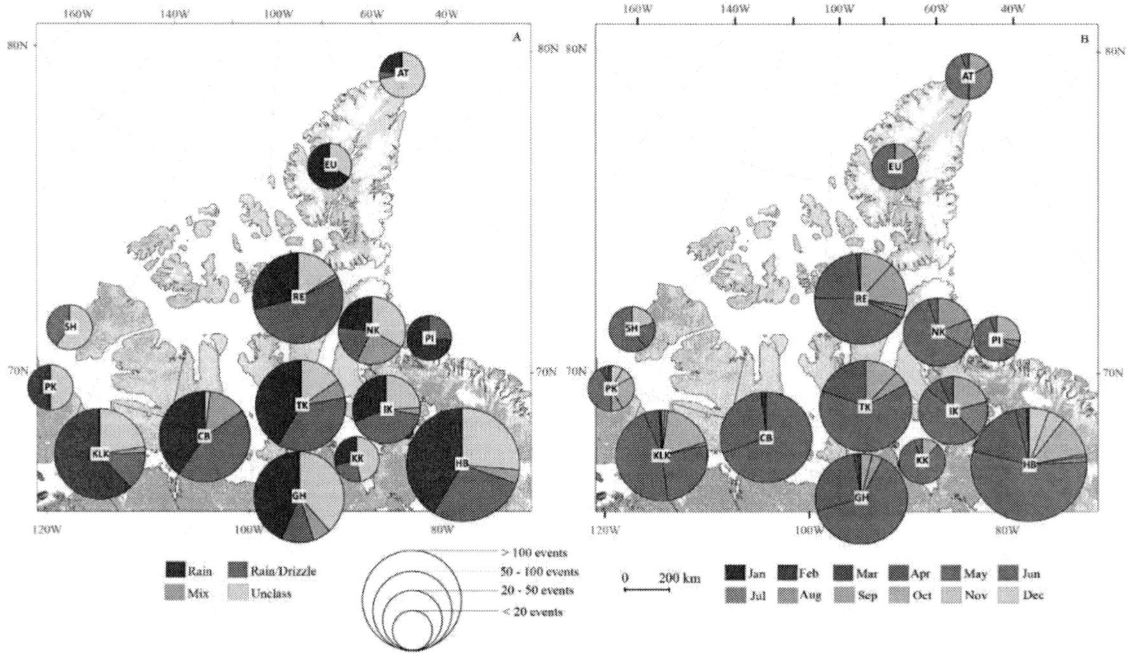

Figure 2. Monthly distribution of ROS events between 1985 and 2013 in the CAA. Pie-chart size represents the number of ROS events and colors represent different months.

We extracted ROS events from the meteorological station database for four periods: fall (~ September 20 to ~ December 20), winter (~ December 20 to ~ March 20), spring (~ March 20 to ~June 20) and summer (~ June 20 to September 20). The temporal occurrences for each of these periods are shown in Figure 3. ROS events are unevenly distributed through the year. Indeed, the spring ROS accounts more than 75% of the total occurrences, while ~14% of the events occur in fall, ~8% during summer and less than 1% in winter. Of relevance, the month of June accounts for more than half of the ROS recorded at all stations throughout the study period. One should note that 13 of the 14 stations received at least one ROS event during spring, whereas five to 11 stations were affected during fall, and one to four during winter. During the fall and winter periods, ROS events have a significant impact on snow metamorphism and stratigraphy with the formation of an ice crust near the surface that affects energy transfer (Colbeck, 1980). Although the metamorphism processes can be observed during spring, the sustained warm air temperatures will promote a transition between the pendular and funicular regimes, leading to deeper water percolation. The modification of snow structure and formation of ice crusts throughout the year can lead to unfavorable grazing conditions for various ungulate species (Sokolov et al., 2016; Ouellet et al. 2017; Langlois et al. 2017).

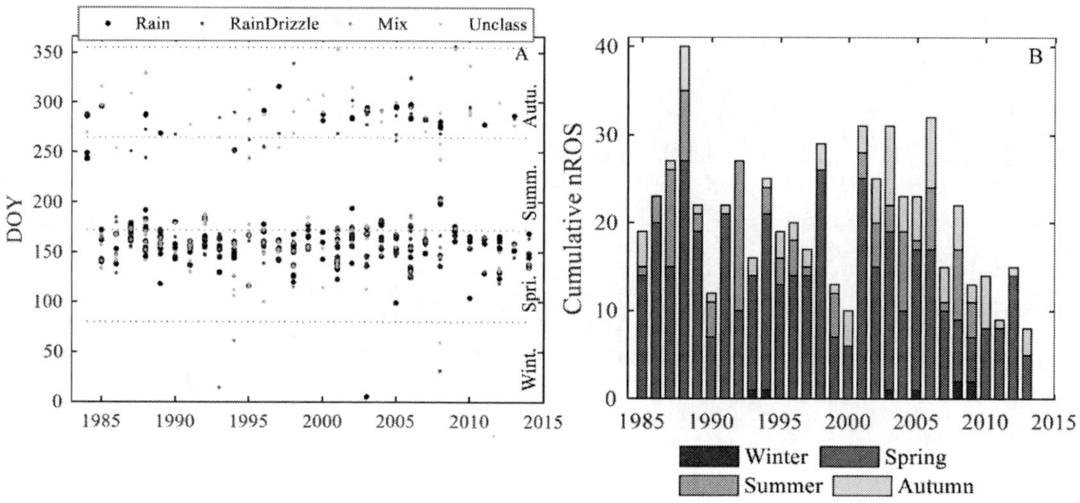

Figure 3. Annual vs. monthly distribution of ROS events (for each class) during the analysis period for all stations combined (A) and seasonal distribution and evolution of ROS events for all stations combined (B).

The results of seasonal trend for fall, winter, spring and summer do not suggest a significant trend due to high year-to-year variability, it is interesting that spring and summer seasons experience fewer ROS events (perhaps linked to a reduced snow cover duration), while an increase in event occurrence is seen during fall and winter. This is relevant for various applications, such as caribou grazing conditions since November-March is the migration period and is a critical time for calf survival.

Interannually ROS distribution by station does not appear to have followed any trend since 1985 (Fig. 4), despite a rise in air temperature. We also investigated potential trends by precipitation class (ROS_{rain}, $ROS_{rain/drizzle}$, ROS_{mix} and $ROS_{Unclass}$) (Fig. 4), and the results did not reach statistical significance. Although an increase in temperature of about 2°C has been clearly visible since 1984 (Fig. 4B), no statistically significant trend can be seen in ROS occurrence. As expected, we see a predominance of low latitude stations (from TK to KLK; i.e., <70N) in the interannually distribution of global ROS (Fig. 4A) and for the ROS_{rain} condition, while we can see a predominance of higher latitude stations (from SH to AT, i.e., >70N) for $ROS_{rain/drizzle}$. For other conditions, no trend can be distinguished (Fig. 4 E, F). Although there is no notable trend, significant variability can be observed where peaks in occurrence can be linked to main northern climate modes. It appears (not shown) that the negative periods of the Arctic Oscillations (as well as the very similar North Atlantic Oscillations, http://www.cpc.noaa.gov) are well synchronized with the observed higher numbers of observed ROS events (>20/yr) between 1984 and 2010. These results agree with Cohen et al., 2015 analysis derived from reanalysis-based study.

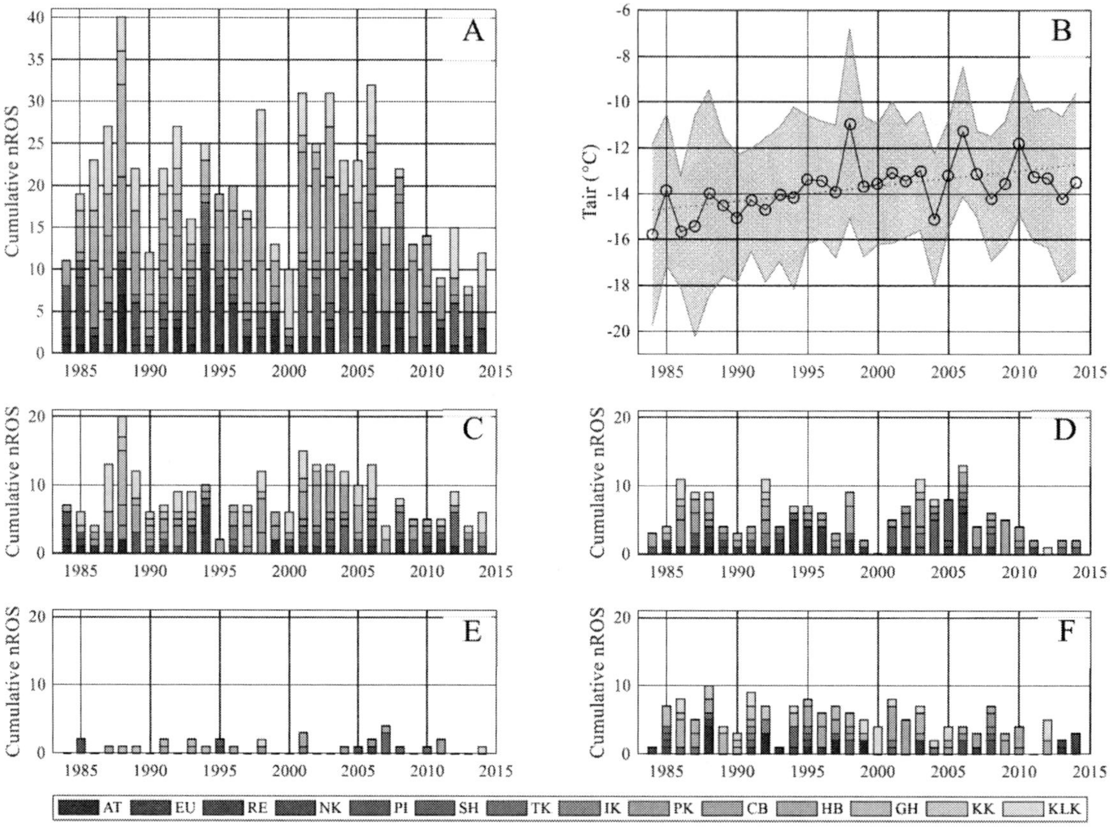

Figure 4. Temporal yearly ROS distribution by station between 1984 and 2014, using all event types (A), for ROS $_{rain}$ (C), ROS $_{rain/drizzle}$ (D), ROS $_{mix}$ (E) and ROS $_{Unclass}$ (F) separately. These figures are compared to winter air temperature (annual mean from ECCC stations) trends for all stations combined during same period (B).

The distinction between different precipitation conditions is important from a microwave radiometry perspective. The presence of rain, drizzle or snow in and on snowpack have a different impact on T_B. In fact, mixed precipitations have a smaller impact on T_B (i.e., GRP), and are thus harder to detect using our algorithm. Furthermore, the relationship between precipitation amount and air temperature did not follow any specific spatial distribution. In Figure 5 we can see this relationship where ROS $_{rain}$ were observed at temperatures above 0°C and the precipitation amounts are concentrated around 2 mm except for KK and NK (Fig. 5A). For the ROS $_{rain/drizzle}$ condition, the values are centered around 0°C, and precipitation amounts remained under 5 mm for all stations (Fig. 5B), while the relationship is more scattered for ROS $_{mix}$ and ROS $_{Unclass}$ conditions with air temperature values varying between -10°C and +5°C and precipitation amount values varying from 0.5 to 2 mm. For ROS $_{Unclass}$, a predominance of small average values of precipitation amounts and air temperatures below 0°C can be seen.

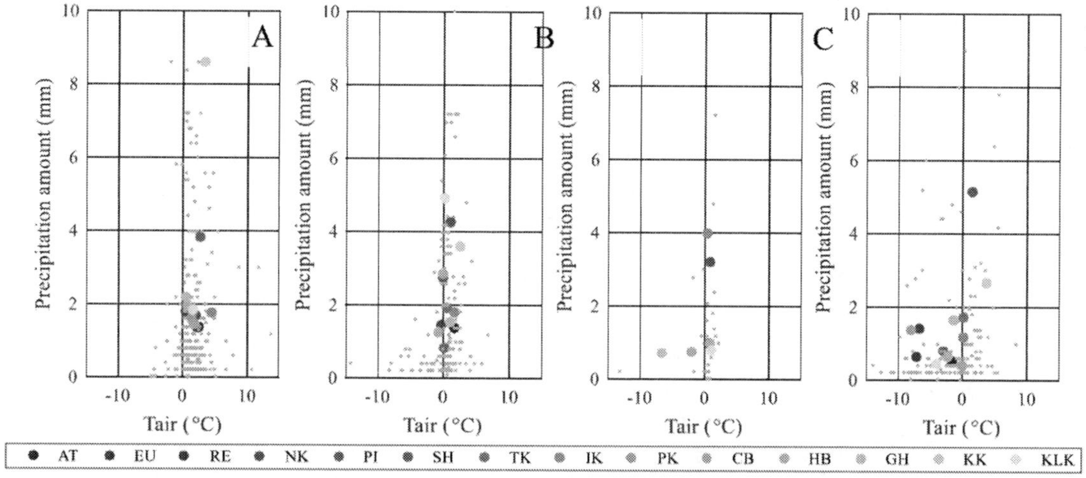

Figure 5. Relationship between air temperature and precipitation amount for rain conditions (A), for rain and drizzle conditions (B), for mix conditions (C) and for unclass conditions (D), between 1984 and 2014. The colored dots correspond to station averages and the grey points are data for all stations combined for different ROS classes.

Figure 6 shows the distribution of precipitation rate (in mm.h^{-1}) for each ROS condition. Most ROS (irrespective of ROS condition) are classified with small precipitation rates (i.e., <1 mm.h^{-1}). It is important to distinguish rate and amount of precipitation. Precipitation rate is derived from the accumulated precipitation and time duration recorded at the meteorological station so that the derived rate value is an hourly approximation (due to a lack of robust data of ROS time duration), and not a direct measurement. Given the rather cold temperatures and low humidity at which ROS occurs in the Arctic, the precipitation rates remain rather small. The precipitation rate itself can be classified into 5 categories (<0.5 mm; 0.5 to 1 mm; 1 to 2 mm; 2 to 5 mm and >5 mm). It is possible to see a predominance of weak precipitation rates (i.e., rate <0.5 mm.h^{-1}), which corresponds to 71.3%, 62.6% and 77.4% of all ROS events for ROS $_{rain}$, ROS $_{rain/drizzle}$ and ROS $_{mix}$, respectively (this statistic is not available for ROS $_{Unclass}$ due to a lack of data).

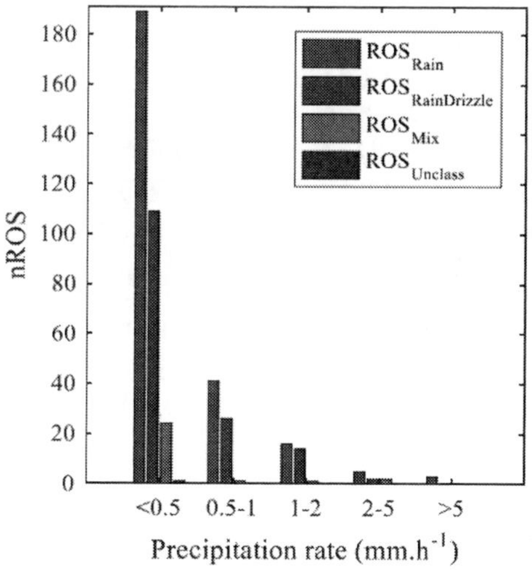

Figure 6. Number of ROS event (nROS) per classes of precipitation rate in mm.h^{-1}, for ROS $_{rain}$, ROS $_{rain/drizzle}$, ROS $_{mix}$ and ROS $_{Unclass}$ conditions, for all stations combined across CAA.

Passive microwave detection using GPR approach

It remains unclear what minimum precipitation amount or rate will lead to a ROS detection using our passive microwave algorithm. As explain in section above, no specific trends are seen and that the omission of small events will remain a source of error. In Figure 7A we thus present the relationship between the GRP threshold and number of ROS detected. One can see a decrease in detected events with a decreasing GRP threshold over a range of -1 to -20 with a more significant decrease between thresholds of 1 (threshold of Dolant et al., 2016) to -5 (*i.e.* when GRP = 1, nROS = 7688 in ascending orbit and 4936 for descending orbit, whereas when GRP = -5, nROS = 289 for ascending orbit and 167 for descending orbit).

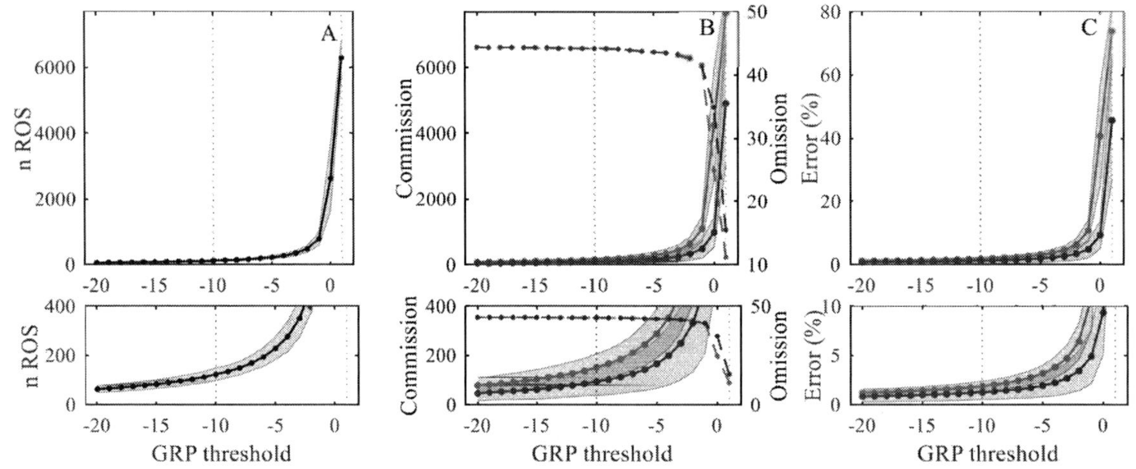

Figure 7. Average number of ROS (nROS) detected as a function of different GRP thresholds for all stations and for both ascending and descending orbits (A). Total number of commission and omission of detection (B) and error evolution (in percent; C) for GRP threshold variation, in subplots B and C, the line corresponds to commission and the dotted line corresponds to omission, the green line corresponds to ascending orbit and the purple line to descending orbit. For all subplots, the grey area represents the standard deviation. For A, B, C the dashed line in grey correspond at threshold of Dolant et al., 2016 and the red dashed line corresponds at new threshold.

We investigated the sensitivity of the detection algorithm to the threshold by comparing omission and commissions over the same range of GRP thresholds. Commissions occur when a ROS is detected by the algorithm, but not observed at the station (commission by the algorithm). Omissions occur when the opposite is observed (ROS seen at station but not detected by the algorithm). The best ROS event detection (perfect match between inventory date and detection date) is achieved when using a low GRP threshold (around -1), but that will lead to a significant increase in commissions (Fig. 7B). Figure 7C also suggests that most of the errors come from commissions and that the total error (omission + commission, see Fig. 7C) stabilizes at -10, where total error is equal to 1.9 for ascending orbit and 1.3 for descending orbit (i.e. omission is around 44 for each orbit and commission is around 152 for ascending orbit and 93 for descending orbit), which becomes the updated threshold for ROS detection. It also agrees with the nROS observed at the stations (Fig. 7A).

DISCUSSION

The main uncertainties using the available ECCC meteorological information are the precipitation timing and amount. First, the satellite passes are fixed, and no information is available on rain amount at a specific time (only daily cumulated values are available). For instance, a satellite can pass overhead during a ROS without detecting it if insufficient water is present to create the T_B reversal between 19 and 37 GHz at horizontal polarization. This would result in an 'omission' from the algorithm. As such, the dataset cannot be used as an absolute reference for validation of the GRP approach, but it provides a collection of rain conditions and T_B responses to evaluate the sensitivity of the threshold (Figs. 7A and 7B). A true validation and adjustment of the approach would require hourly precipitation phase data, which will be available in Cambridge Bay in 2018 (disdrometer installation planned). Even though these uncertainties are present, this work allows a good framework of ROS detection measurement, and thus a good overview of this phenomenon throughout the CAA area. Other studies will evaluate the link between meteorological conditions, ROS intensity and impact intensity in snow structure and in environment. Nonetheless, our results highlight the best compromise for detection accuracy, where a threshold of -10 suggests a good equilibrium between omissions and commissions. One could reduce the threshold to increase the number of perfect detection matches, however this would translate into an increase in commissions, thereby reducing the overall accuracy. Moreover, this threshold could be adjustable for different applications and different levels of application (e.g., increasing the threshold to -5, where here more than 200 ROS could be detected, for civil security in Inuit communities (Berkes and Jolly, 2001)).

The periods for which visual observations are available is another source of uncertainty. However, the datasets included several gaps in these periods, occasionally with no observations for several consecutive days, which meant that the algorithm had no reference data. Consequently, it is hard to conclude on spatial and/or temporal event occurrence trends using this dataset.

CONCLUSION

This study presented ROS occurrences at 14 stations across the Arctic between 1984 and 2014 to evaluate the robustness of the GRP approach developed by Dolant et al., 2016 in an Arctic context. The basic statistical analysis highlighted a total of 603 ROS events throughout the study period (1984-2014), dominated by event occurrences at lower latitude stations. Three periods were studied separately and results show that despite no significant trends in the cumulated yearly occurrence, an increase is seen in the fall and winter ROS events.

We used the ROS inventory dataset to evaluate the performance and sensitivity of the GRP threshold developed by Dolant et al. (2016). Results suggest using a different threshold to differentiate between pure rain events and mixed precipitations with GRP thresholds of -15 and -10, respectively. Those thresholds also appear to match with the lowest detection error computed with omissions and commissions. Intuitively, the next step is to compute binary images of detected ROS across the CAA, and evaluate yearly anomalies in event occurrence (e.g., links to El Nino and Arctic Oscillation). Furthermore, the remote sensing product will also allow a better understanding of the various processes that can trigger an ROS. For instance, binary maps of event occurrences could be compared to sea proximity and polynya occurrence and evaluate future spatio-temporal trends of ROS across the Arctic.

REFERENCES

Berkes F, Jolly D. 2001, Adapting to climate change: social-ecological resilience in a Canadian western Arctic community. *Conservation Ecology*, **5**: 18. doi: 10.5751/ES-00342-050218.

Bokhorst S, Pedersen SH, Brucker L, *et al.* 2016. Changing Arctic snow cover: A review of recent developments and assessment of future needs for observations, modelling and impacts. *Ambio*, **45**: 516-537. doi: 10.1007/s13280-016-0770.

Brown RD, Braaten RO. 1998. Spatial and temporal variability of Canadian monthly snow depths, 1946-1995. *Atmosphere-Ocean,* **36**: 37-45.

Brown RD, Mote PW. 2009. The response of Northern Hemisphere snow cover to a changing climate. *Journal of Climate*, **22**: 2124-2145. doi: 10.1175/2008JCLI2665.1.

Brown R, Derksen C, Wang L. 2010. A multi-data set analysis of variability and change in Arctic spring snow cover extent, 1967–2008. *Journal of Geophysical Research*, **115**: D16111. doi:10.1029/2010JD013975.

Chen W, Russell DE, Gunn A, Croft B, Chen WR, Fernandes R, Zhao H, Li J, Zhang Y, Koehler K, Olthof I, Fraser RH, Leblanc SG, Henry GR, White RG, Finstad GL. 2013. Monitoring

habitat condition changes during winter and pre-calving migration for Bathurst Caribou in northern Canada. *Biodiversity*, **14**: 36-44.
Cohen J, Ye H, Jones J. 2015. Trends and variability in rain-on-snow events, *Geophysical Research Letters.*, **42.** doi:10.1002/2015GL065320.
Colbeck S. 1980. Thermodynamics of snow metamorphism due to variations in curvature. *Journal of Glaciology,* **26**: 291-301.
Colton MC, Poe GA. 1999. Intersensor Calibration of DMSP SSM/I's: F-8 to F-14, 1987-1997. *IEEE Transactions on Geoscience and Remote Sensing,* **37**: 418-439.
Comiso JC. 2014. Sea Ice Concentration and Extent. *Encyclopedia of Remote Sensing,* 727-743.
Cullather RI, Lim Y-K, Boisvert LN, Brucker L, Lee JN, Nowicki SMJ. 2016. Analysis of the warmest Arctic winter, 2015-2016. *Geophysical Research Letters*, **43**. doi:10.1002/2016GL071228.
Derksen C, Smith SL, Sharp M, Brown L, Howell S, Copland L, Mueller DR, Gauthier Y, Fletcher CG, Tivy A, Bernier M, Bourgeois J, Brown R, Burn CR, Duguay C, Kushner P, Langlois A, Lewkowicz AG, Royer A, Walker A. 2012. Variability and change in the Canadian cryosphere. *Climatic change,* **11**: 59-88.
Derksen C, Brown R. 2012. Spring snow cover extent reductions in the 2008–2012 period exceeding climate model projections. *Geophysical Research Letter,* **39**, 6p.
Dethloff K, Rinke A, Benkel A, Køltzow M, Sokolova E, Kumar Saha S, Handorf D, Dorn W, Rockel B, Von Storch H, Haugen JE, Røed LP, Roeckner E, Christensen JH, Stendel M. 2006. A dynamical link between the Arctic and the global climate system. *Geophysical Research Letters*, **33**: L03703. doi:10.1029/2005GL025245.
Dolant C, Langlois A, Montpetit B, Brucker L, Roy A, Royer A. 2016. Development of a rain-on-snow detection algorithm using passive microwave radiometry. *Hydrological Processes,* **30**: 3184–3196.
Domine F, Barrere M, Sarrazin D. 2016. Seasonal evolution of the effective thermal conductivity of the snow and the soil in high Arctic herb tundra at Bylot Island, Canada. *Cryosphere*, **10**: 2573-2588.
Grenfell TC, Putkonen J. 2008. A method for the detection of the severe rain-on-snow event on Banks Island, using passive microwave remote sensing. *Water Resources Research*, **44**: W03425. doi: 10.1029/2007WR005929.
Holland MM, Serreze MC, Stroeve J, 2010. The sea ice mass budget of the Arctic and its future change as simulated by coupled climate models. *Climate Dynamics*, **34**: 185–200.
IPCC. 2013. Climate Change 2013: The Physical Science Basis. Contribution of Working Group I to the Fifth Assessment Report of the Intergovernmental Panel on Climate Change [Stocker TF, Qin D, Plattner G-K, Tignor M, Allen SK, Boschung J, Nauels A, Xia Y, Bex V, Midgley PM. (eds.)]. *Cambridge University Press, Cambridge, United Kingdom and New York, NY, USA*, 1535 pp.
IPCC. 2014. Climate Change 2014: Impacts, Adaptation, and Vulnerability. Part A: Global and Sectoral Aspects. Contribution of Working Group II to the Fifth Assessment Report of the Intergovernmental Panel on Climate Change [Field CB, Barros VR, Dokken DJ, Mach KJ, Mastrandrea MD, Bilir TE, Chatterjee M, Ebi KL, Estrada YO, Genova RC, Girma B, Kissel ES, Levy AN, MacCracken S, Mastrandrea PR, White LL (eds.)]. *Cambridge University Press, Cambridge, United Kingdom and New York, NY, USA*, 1132 pp.
IPCC. 2014. Climate Change 2014: Impacts, Adaptation, and Vulnerability. Part B: Regional Aspects. Contribution of Working Group II to the Fifth Assessment Report of the Intergovernmental Panel on Climate Change [Barros VR, Field CB, Dokken DJ, Mastrandrea MD, Mach KJ, Bilir TE, Chatterjee M, Ebi KL, Estrada YO, Genova RC, Girma B, Kissel ES, Levy AN, MacCracken S, Mastrandrea PR, White LL (eds.)]. *Cambridge University Press, Cambridge, United Kingdom and New York, NY, USA*, pp. 688.
IPCC. 2014. Climate Change 2014: Synthesis Report. Contribution of Working Groups I, II and III to the Fifth Assessment Report of the Intergovernmental Panel on Climate Change [Core Writing Team, Pachauri RK, Meyer LA (eds.)]. IPCC, Geneva, Switzerland, 151 pp.
Johnson CA, Neave E, Blukacz-Richards A, Banks SN, Quesnelles PE. 2016. Knowledge assessment (community and scientific) to inform the identification of critical habitat for Peary

caribou, *Rangifer tarandus pearyi*, in the Canadian Arctic. *Environment and Climate Change Canada, Science and Technology, Ottawa, Ontario, Canada.*
Klos PZ, Link TE, Abatzoglou JT. 2014. Extent of the rain-snow transition zone in the western U.S. under historic and projected climate. *Geophysical Research Letters*, **41**: 4560–4568.
Knowles K, Njoku EG, Armstrong R, Brodzik M. 2000. Nimbus-7 SMMR Pathfinder Daily EASE-Grid Brightness Temperatures, Version 1. *Boulder, Colorado USA, NASA National Snow and Ice Data Center Distributed Active Archive Center.*
Langlois A, Johnson CA, Montpetit B, Royer A, Blukacz-Richards EA, Neave E, Dolant C, Roy A, Arhonditsis G, Kin DK, Kaluskar S, Brucker L. 2017. Detection of rain-on-now (ROS) events and ice layer formation using passive microwave radiometry: A context for Peary caribou habitat in the Canadian Arctic. *Remote Sensing of Environment*, **189**: 84-95.
Liebe H. 1989. MPM-An atmospheric millimeter-wave propagation model. *International Journal of Infrared and Millimeter Waves*, **10**: 631-650.
Liston GE, Hiemstra CA. 2011. The Changing Cryosphere: Pan-Arctic Snow Trends (1979–2009). *American Meteorological Society.*
L'hôte Y, Chevallier P, Coudrain A, Lejeune Y, Etchevers P. 2005. Relationship between precipitation phase and air temperature: comparison between the Bolivian Andes and the Swiss Alps. *Hydrological Sciences Journal*, **50**: 989-997.
Mazurkiewicz AB, Callery DG, McDonnell JJ. 2008. Assessing the controls of the snow energy balance and water available for runoff in a rain-on-snow environment. *Journal of Hydrology*, **354**: 1-14.
Mesinger F, DiMego G, Kalnay E, Mitchell K, Shafran PC, Ebisuzaki W, Jović D, Woollen J, Rogers E, Berbery EH, Ek MB, Fan Y, Grumbine R, Higgins W, Li H, Lin Y, Manikin G, Parrish D, Shi W. 2006. North American Regional Reanalysis. *Bulletin of the American Meteorological Society*, **87**: 343-360.
Montpetit B, Royer A, Roy A, Langlois A, Derksen C. 2013. Snow microwave emission modeling of ice lenses within the snowpack using the microwave emission model for layered snowpack (MEMLS). *IEEE Transactions on Geoscience and Remote Sensing.*
Montpetit B. 2015. Analyse de la modélisation de l'émission multifréquence micro-onde des sols et de la neige, incluant les croutes de glace à l'aide du modèle Microwave Emission Model of Layered Snowpacks (MEMLS). *Ph.D. thesis, Université de Sherbrooke, Département de Géomatique Appliquée,* May 2015, 188p.
Ouellet F, Langlois A, Johnson CA, Richards A, Royer A. 2017. Spatialization of the SNOWPACK Snow model in the Canadian Arctic for Peary Caribou Winter Grazing Condition Assessment. *Physical Geography*, **38**: 143-158.
Pradhanang SM, Frei A, Zion MS, Schneiderman EM, Steenhuis TS, Pierson D. 2012. Analysis of Rain-on-snow Runoff Events in New York. *69th Eastern Snow Conference, Frost Valley YMCA, Claryville, New York, USA.*
Pradhanang SM, Frei A, Zion M, Schneiderman EM, Steenhuis TS, Pierson D. 2013. Rain-on-snow runoff events in New York. *Hydrological Processes*, **27**: 3035-3049.
Pulliainen J, Grandell T, Hallikainen J. 1999. Hut snow emission model and its applicability to snow water equivalent retrieval. *IEEE Transactions on Geoscience and Remote Sensing*, **37**: 1378-1390.
Putkonen J. 1998. Soil thermal properties and heat transfer processes near Ny Alesund, north western Spitsbergen, Svalbard. *Polar Research*, **17**: 165–179.
Putkonen J, Roe G. 2003. Rain-on-snow events impact soil temperatures and affect ungulate survival. *Geophysical Research Letters*, **30**: 1188.
Rennert KJ, Roe G, Putkonen J, Bitz CM. 2009. Soil Thermal and Ecological Impacts of Rain on Snow Events in the Circumpolar arctic. *Journal of Climate*, **22**: 2302-2315.
Romanovsky VE, Smith SL, Christiansen HH. 2010. Permafrost Thermal State in the Polar Northern Hemisphere during the International Polar Year 2007–2009: A Synthesis. *Permafrost and Periglacial Processes*, **21**: 106–116.
Roy A, Royer A, Wigneron JP, Langlois A, Bergeron J, Cliche P. 2012. A simple parameterization for boreal forest radiative transfer model at microwave frequencies. *Remote Sensing of Environment,* **124**: 371-383.

Serreze MC, Barrett AP, Stroeve JC, Kinding DN, Holland MM. 2009. The emergence of surface-based Arctic amplification. *The Cryosphere*, **3**: 11-19.

Serreze MC, Barry RG. 2011. Processes and impacts of Arctic amplification: A research synthesis. *Global and Planetary Change*, **77**: 85-96.

Steele M, Ermold W, Zhang J. 2008. Arctic Ocean surface warming trends over the past 100 years, *Geophysical Research Letters*, **35**: L02614. doi:10.1029/2007GL031651.

Sokolov AA, Sokolova NA, Ims RA, Brucker L, Ehrich D. 2016. Emergent Rainy Winter Warm Spells May Promote Boreal Predator Expansion into the Arctic. *Arctic*, **69**: 121. doi:10.14430/arctic4559.

Trenberth, KE, Jones PD, Ambenje P, Bojariu R, Easterling D, Klein Tank A, Parker D, Rahimzadeh F, Renwick JA, Rusticucci M, Soden B, Zhai P. 2007. Chapter 3, Observations: Surface and Atmospheric Climate Change. In: Climate Change 2007: The Physical Science Basis. Contribution of Working Group I to the Fourth Assessment Report of the Intergovernmental Panel on Climate Change [Solomon S, Qin D, Manning M, Chen Z, Marquis M, Averyt KB, Tignor M, Miller HL (eds.)]. *Cambridge University Press, Cambridge, United Kingdom and New York, NY, USA.*

Trenberth KE. 2011. Changes in precipitation with climate change. *Climate Research*, **47**: 123–138.

Wentz FJ, Spencer RW. 1988. SSM/I Rain Retrievals within a Unified All-Weather Ocean Algorithm. *American Meteorological Society,* **55**: 1613-1627.

Wentz FJ. 1997. A well-calibrated ocean algorithm for special sensor microwave/imager. *Journal of Geophysical Research,* **10**: 8703-8718.

Wentz FJ. 2013. SSM/I Version-7 Calibration Report. *Remote Sensing Systems Technical Report,* 46p.

Winton M. 2006. Does the Arctic sea ice have a tipping point? *Geophysical Research Letters*, **33**: 5p.

Campbell Scientific Canada Award for Best Student Paper Demonstrating Innovative Use of Technology

Tracking Changes in Iceberg Calving Events and Characteristics from Trinity and Wykeham Glaciers, SE Ellesmere Island, Canada

ABIGAIL DALTON[1] AND LUKE COPLAND[1]

ABSTRACT

The Canadian Ice Service (CIS) produces charts to identify iceberg presence in Canadian waters for navigational purposes, but there is uncertainty surrounding the origin of these features. Previous work has been conducted on the location of the main tidewater glaciers in Canada and the volumes of ice that they discharge to the ocean, but little research has been done on the processes controlling iceberg production and drift patterns in Canadian Arctic waters. Recent and rapid changes have been observed in the ice discharge patterns of glaciers in this region. For example, Trinity and Wykeham glaciers contributed ~62% of total iceberg discharge from the Canadian Arctic Archipelago in 2016, compared to ~22% in 2000. Given these changes, an important question is whether there is a relationship between changing sea ice conditions (e.g., thickness, extent, freeze up dates, break up dates) and iceberg production from these glaciers. Using SAR and optical imagery to identify iceberg calving events between 1997 and 2015, iceberg calving events are classified in two ways: through the identification of iceberg plumes, and through the measurement of glacier terminus changes. Often when a calving event occurs a plume of glacial ice debris is also produced at the terminus and is clearly visible in SAR imagery. In some cases, such as when a large event composed of a single iceberg rather than many small pieces of ice, a plume is not always produced making it difficult to detect in SAR imagery. Glacier area loss detected in optical imagery data is compared to the plume event data from SAR imagery between 1997 and 2015 to determine whether a relationship exists between glacier area loss and plume size. To understand where these icebergs drift to, how they deteriorate and the time scale of these processes, a series of satellite tracking beacons were deployed in summer 2016. These were helicopter-deployed from aboard the CCGS Amundsen, and provide near real-time (hourly) information concerning the movement of 13 icebergs and ice islands within Baffin Bay. Initial results show that, to date, the most active iceberg has drifted over 3200km south through Baffin Bay. Some of the icebergs have also exhibited a spiraling pattern as they drift west across Baffin Bay and are influenced by ocean currents and tides. Results from this work provide information about patterns of iceberg movement, including common areas where icebergs become grounded in relation to bathymetry.

Keywords: Icebergs, Remote Sensing, Glacier Dynamics, Iceberg Drift, Canadian Arctic

[1] University of Ottawa, Department of Geography, Environment and Geomatics, 60 University Private, SMD 043 Ottawa, ON, K1N 8Z4

GLACIERS, ICEBERGS AND ICE ISLANDS

Background and Objectives

Tidewater glaciers drain glaciers, ice caps and ice sheets and terminate into the ocean where they discharge through the calving of icebergs and ice islands (large tabular icebergs). The Canadian Ice Service (CIS) produces charts to identify the presence of icebergs but has little knowledge about the sources and sinks of icebergs in Canadian waters. It is important to understand where these icebergs and ice islands originate, where they drift, how they deteriorate and the time scale of these processes. Trinity and Wykeham Glaciers on SE Ellesmere Island have increased iceberg production from 22% of total discharge from the CAA (Canadian Arctic Archipelago) in 2000 to 62% in 2016. They are the only two glaciers in the Queen Elizabeth Islands (QEI) to have shown consistent acceleration between 1999 and 2015 making it an area of significance for the study of ice discharge into Canadian Waters (Van Wychen et al., 2016). Operations during this leg will address the following gaps in knowledge surrounding the production and movement of icebergs and ice islands in Canadian waters:

- Which tidewater glaciers are the sources of icebergs and ice islands in Canadian waters and where do they drift?
- Are there changes in the size, shape or timing of iceberg production in the recent past and is this linked to glacier dynamics?
- Do sea ice conditions impact the production of icebergs at the termini of tidewater glaciers?

Iceberg Plume Detection

Trinity and Wykeham Glaciers were chosen to be the focus of this study due to recent and rapid changes in their velocities and ice discharge identified by Van Wychen et al. (2016). Through plume analysis, they were found to be the most active iceberg plume producing glaciers in the study with 139 and 134 plumes detected, respectively. Trinity Glacier was found to be the most active glacier in this study with regards to iceberg plume production (Figure 1). Most of the plume events detected occurred during the open water season or immediately before/after with fewer events occurring when sea ice was present. Very few plume events were detected between January and June for all years, with the only winter events being seen in recent years (2013-2015). Most of the plumes produced by Trinity Glacier were size 1 or size 2 in magnitude. Trinity Glacier produced the most size 3 plume events of all glaciers in the study with 11. Similar to Trinity Glacier, most of the iceberg plume events produced by Wykeham Glacier (Figure 2) were detected during the open water season however more events were detected when sea ice was present (approximately December to May). All of these events were smaller in size (size 1 or 2) compared to the events that occurred during the open water season. Two size 3 events and one size 4 event were produced by Wykeham Glacier however these were only detected in recent years, between 2011 and 2014. Sea ice did not break out of Trinity Fiord in 1997 and no iceberg plume events were detected from either Trinity or Wykeham Glaciers during that year. In summary, when sea ice is present, fewer iceberg calving events occur. Once the sea ice begins to break out of the fiord there is an increase in the size and frequency of iceberg calving events.

Iceberg Beacon Deployment

Between July 31, 2016 and August 14, 2016 a total of 13 beacons were deployed on icebergs and ice islands in Baffin Bay. Through a contract with Environment and Climate Change Canada (ECCC) six CALIB beacons were deployed onto icebergs and ice islands within Baffin Bay and Nares Strait (Table 1). Two beacons were deployed onto targets chosen by CIS and the remaining four targets were chosen based on size, location and whether they were likely to drift (Figure 3).

Five of six beacons have since successfully transmitted data remotely. One of the beacons deployed (ID#: 300234011758690) is suspected to have had issues with the battery and has not transmitted any data since being deployed.

Seven additional beacons were deployed containing iridium GPS receivers (RockStar), batteries and solar panels (Figure 4). Three of these beacons were deployed onto icebergs/ice islands within Baffin Bay and four were deployed within Trinity Fiord to track movement of icebergs produced by Trinity Glacier within and out of the fiord (Table 2; Figure 6). Positions of these seven beacons will be viewed on www.core.rock7.com to monitor movement and identify drift patterns of icebergs around Baffin Bay. An additional 14 beacons were deployed onto icebergs and ice islands in 2017. Nine of these beacons were of a similar design to those in Figure 5 however they were updated with a larger battery and the addition of a solar regulator. The remaining five beacons were deployed on behalf of CIS.

Table 1. CIS beacon deployment summary.

Asset Name	*Deployment Date*	*Deployment Time (UTC)*	*Latitude (start position)*	*Longitude (start position)*	*Deployment Location*	*Notes*
ICALIB (4701652-8690)	02-Aug-16	11:30 AM	72.06178	-73.31847	Baffin Bay	Buchan Gulf Iceberg. **Faulty Device - Not working**
ICALIB (4701653-3030)	02-Aug-16	12:20 PM	71.88914	-72.25875	Baffin Bay	Unk_S1 Ice Island
ICALIB (4701654-5450)	09-Aug-16	11:35 UTC	76.30836	-74.81636	Baffin Bay	S Nares Strait Ice Island
ICALIB (4701655-1040)	09-Aug-16	12:45 UTC	76.84956	-75.88631	Baffin Bay	East of Manson Icefield Ice Island
ICALIB (4701657-3450)	14-Aug-16	11:31 AM	79.74294	-64.96186	Baffin Bay	Humboldt Glacier Ice Island
ICALIB (4701656-8060)	14-Aug-16	11:49 AM	79.83578	-67.35725	Baffin Bay	Kane Basin Ice Island

Table 2. RockStar beacon deployment summary.

Unit #	**Deployment Date**	**Deployment Time (UTC)**	**Latitude (start position)**	**Longitude (start position)**	**Deployment Location**
3655	31-Jul-16	18:37	70°45'46.76"	67°51'26.50"	SE of Sam Ford Fiord
3534	06-Aug-16	18:34	76°11'02.29"	69°55'31.80"	SW of Thule
3651	06-Aug-16	18:58	76°35'15.51"	71°35'00.91"	W of Thule
20781	10-Aug-16	17:54	77°57'01.57"	78°31'25.99"	Trinity Terminus
3635	10-Aug-16	17:59	77°56'07.96"	78°08'41.42"	Trinity Island
20785	10-Aug-16	10:22	77°56'30.47"	77°55'50.91"	Trinity Mid Fiord
20784	10-Aug-16	10:28	77°54'17.19"	77°29'8.83"	Trinity Outer Fiord

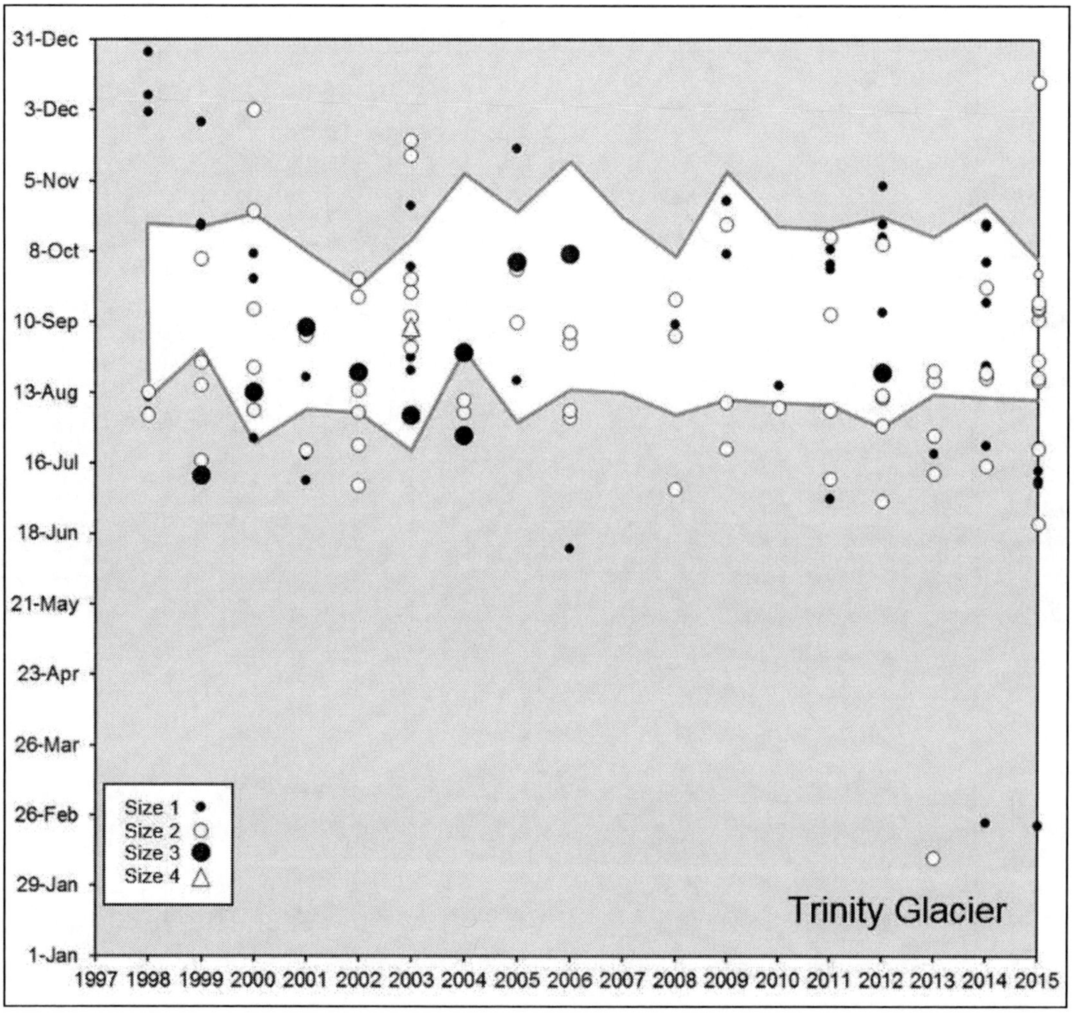

Figure 1. Temporal variability of iceberg plumes produced by Trinity Glacier between 1997 and 2015. Grey area represents the presence of landfast sea ice and white area represents absence of landfast sea ice.

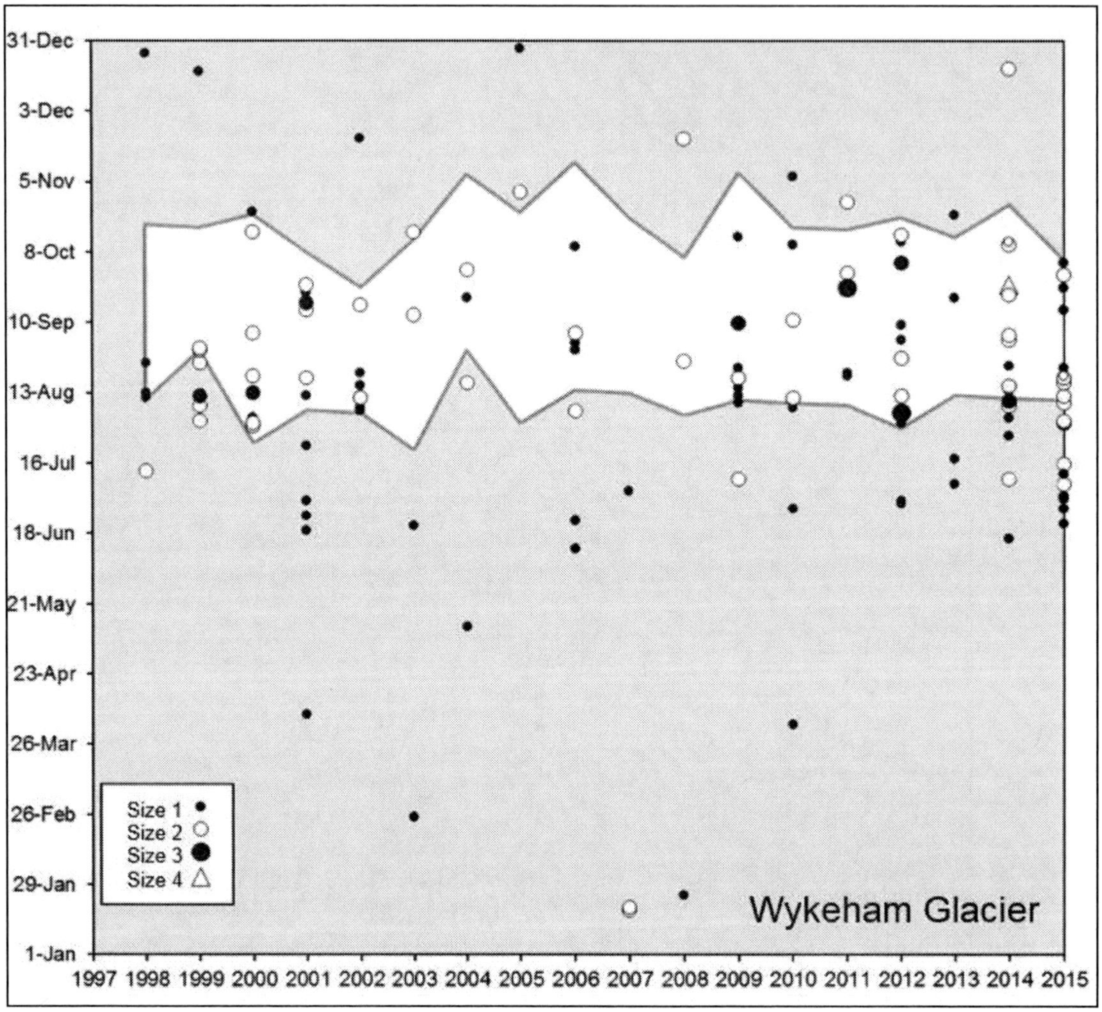

Figure 2. Temporal variability of iceberg plumes produced by Wykeham Glacier between 1997 and 2015. Grey area represents the presence of landfast sea ice and white area represents absence of landfast sea ice.

Figure 3. Example of placement of CIS beacon. Deployed on iceberg near Buchan Gulf (72.06178, -73.31847), August 2, 2016. Photo courtesy of Luke Copland.

Figure 4. Example of placement of beacon on ice island. Deployed SE of Sam Fiord, July 31, 2016. Photo courtesy of Abby Dalton.

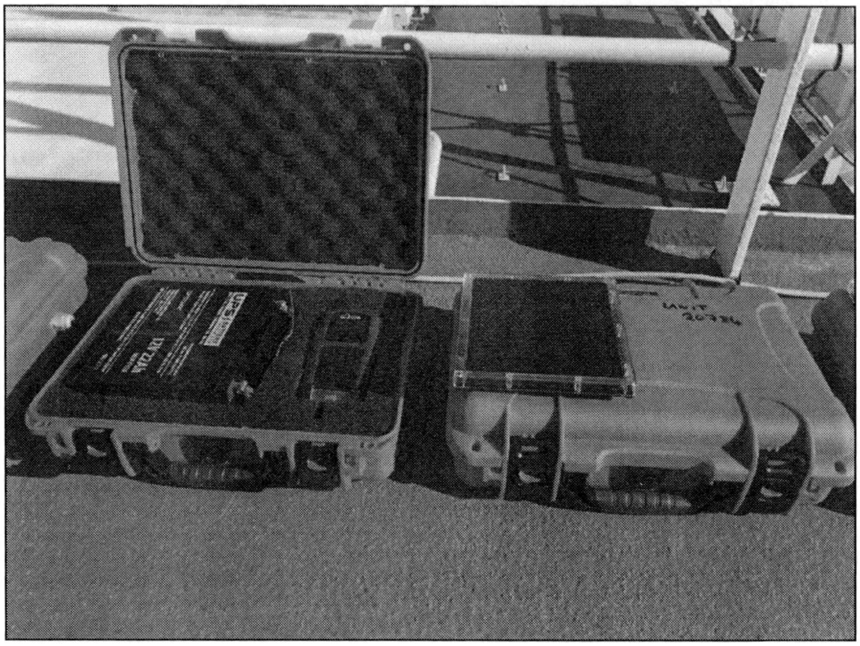

Figure 5. Iceberg tracking beacons built by Abby Dalton and Luke Copland. Three of these beacons were deployed onto icebergs in Baffin Bay and four were deployed onto icebergs within Trinity Fiord to monitor their movement. GPS trackers transmit their position using the Iridium satellite network and can be programmed remotely to modify transmission frequency according to activity (e.g. less frequent transmission during winter when they are frozen in sea ice and have minimal movement). Photo: Luke Copland.

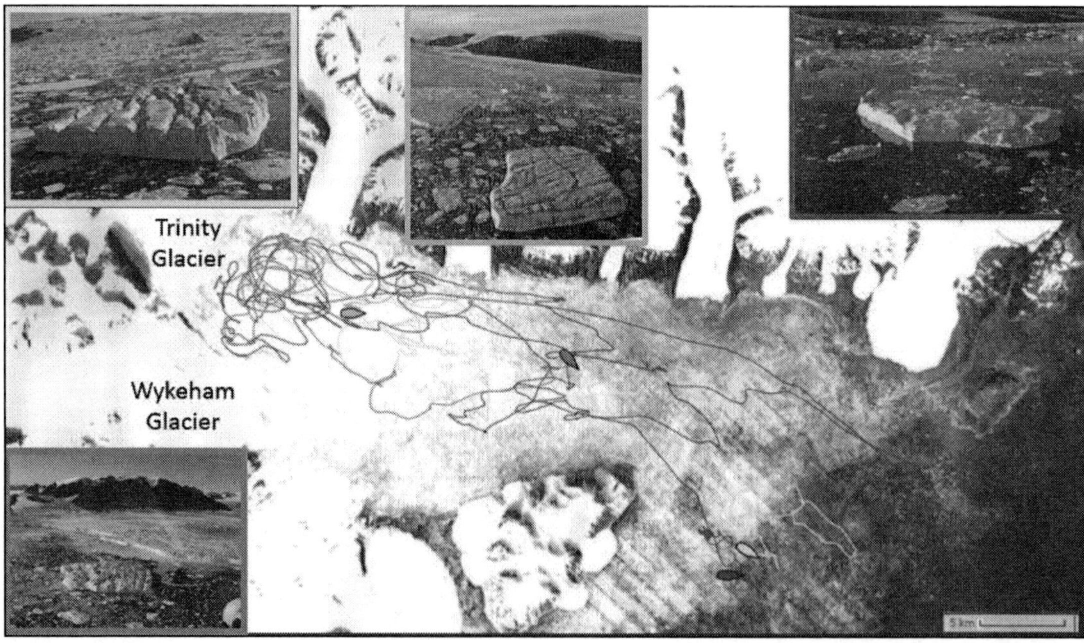

Figure 6. Drift tracks of icebergs tagged with beacons between August and December 2016. One of the icebergs (purple track) travelled approximately 25km in one day on September 13, 2016.

REFERENCES

Van Wychen W, Davis J, Burgess DO, Copland L, Gray L, Sharp M, Mortimer C. 2016. Characterizing interannual variability of glacier dynamics and dynamic discharge (1999-2015) for the ice masses of Ellesmere and Axel Heiberg Islands, Nunavut, Canada. *Journal of Geophysical Research – Earth Surface*, **121**. doi: 10.1002/2015JF003708.

David Miller Award for Best Student Poster

Sensitivity of Advanced Microwave Scanning Radiometer (AMSR-E) Based Snow Water Equivalent (SWE) Retrievals to Rain-on-Snow (ROS) Events in Polar Regions

ELIZABETH RYAN[1,3], LUDOVIC BRUCKER[2,3], AND BARTON A. FORMAN[1]

ABSTRACT

During the winter month, the occurrence of Rain-on-Snow (ROS) events can impact snow stratigraphy via generation of large-scale ice crusts, either on or within the snowpack. The formation of such layers can significantly alter the electromagnetic response of the snowpack. Observed brightness temperatures (Tb) at different microwave frequencies and polarizations depend on various snow properties. The presence of liquid water (rather than solid ice) induces a shift in the electromagnetic response from a dry scatter to a wet emitter and thus yields a larger Tb than a typical dry snowpack would. Considering that various SWE retrievals heavily rely on remotely-sensed observations of passive microwave Tb emission, ROS events could significantly impact SWE estimation. This study aims to discern a linkage between local SWE fluctuations and ROS events to determine if ROS events should be flagged in passive microwave based SWE retrievals at locations where there may be potential error.

Using the National Snow and Ice Data Center (NSIDC) AMSR-E/Aqua L3 Global SWE EASE-Grid product, changes in SWE before, during, and after a known or estimated ROS event were calculated and compared to the SWE time series of other ROS events. Due to the lack of documented historical ROS events, three different approaches were employed in this study to detect ROS events: One approach is solely based on microwave Tb observations, one from in-situ measurements using NOAA's Global Surface Summary of the Day (GSOD) ground-based weather station network, and one using the output of an advanced land surface model. Therefore, this research has the potential to further explore the feasibility of various ROS detection approaches, in addition to, help inform passive microwave SWE retrievals or snow cover properties.

[1] University of Maryland, College Park, Department of Civil and Environmental Engineering
[2] NASA Goddard Space Flight Center, Cryospheric Science Laboratory, code 615
[3] Universities Space Research Association, GESTAR

Oral Session #2: In Situ Snow Measurements

Inexpensive In-Situ Snow Pack Sensors for Temperature, Density, and Grain Size: A First Look at the First Season

ROGER DE ROO,[1] ERIC HAENGEL,[2] STEVEN ROGACKI[3], CHANDLER EKINS[4], AND SEYEDMOHAMMAD MOUSAVI[5]

ABSTRACT

A suite of small, battery operated devices for implantation in a snow pack has made its first season of measurements in the Winter of 2017. They measure and log sensor outputs related to snow pack parameters of temperature, density, moisture and grain size. The temperature is provided by an electronic thermometer; snow density and moisture affect an open resonant circuit operating near 900MHz; grain size and density affect the scattering of an optical link operating at 880nm.

Ten units were deployed at the Mesa Lakes Ranger Station on Grand Mesa, CO as part of the Local Scale Observation Site (LSOS) of NASA's Snow Experiment (SnowEx). The LSOS hosted a micro-meteorological station for the same duration as the longest deployed snow sensor units. During the SnowEx Intensive Observation Period (IOP), from 06 to 24 Feb 2017, the LSOS was the site of snow pit excavations. The units were deployed in pairs on 01 Oct 2016 (on the ground), twice during a snowstorm on 16 Dec 2016, and twice on 20 Feb 2017. In all cases, the units were placed by hand on the top of the snowpack, with its sensor face vertical and facing north. The units deployed in Feb were retrieved on 25 Feb, at the end of the IOP.

Laboratory-based calibrations have been applied to the data, and the resulting snowpack parameters are compared to available ground truth collected adjacent to them.

Keywords: Snow, temperature, density, grain size, time-series.

INTRODUCTION

The quantification of the accumulation of snow on the ground is notoriously difficult to achieve. One reason for the slow progress of remote sensing techniques for the measurement of snow on the ground is the fact that even obtaining ground truth is difficult. This document presents a means of collecting time series of point measurements of parameters of snow packs, using a design for a small, low-power, wireless sensor module that logs information about the temperature, density, wetness and snow optical diameter that is local to the data logger.

---

[1] Climate and Space Science and Engineering, University of Michgan, Ann Arbor. email: deroo@umich.edu

[2] Leidos Corporation, Arlington, VA.

[3] Space Physics Research Laboratory, University of Michigan, Ann Arbor.

[4] Electrical Engineering and Computer Science, University of Michigan, Ann Arbor

[5] Electrical Engineering and Computer Science, University of Michigan, Ann Arbor

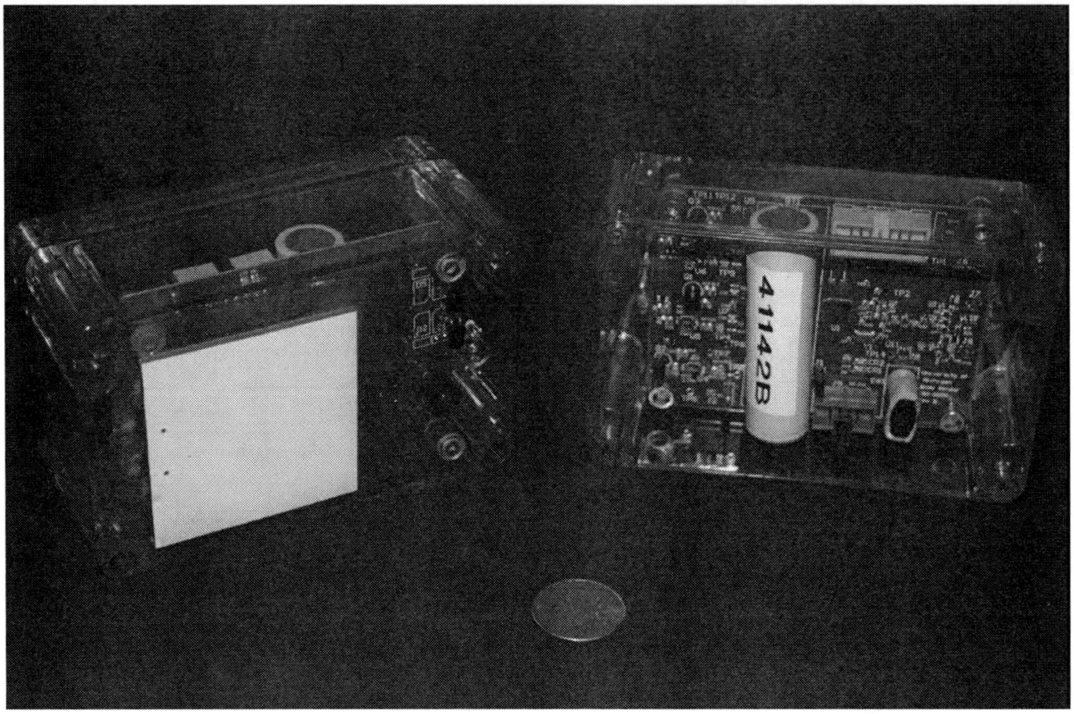

Figure 1. Two completed snow sensor modules that measure snow pack temperature, density, moisture, and grain size. The coin is a US quarter dollar. The unit on the left has the sensor face towards the camera; the unit on the right has its user interface towards the camera and the lid removed for clarity.

This information is the most important for modeling the brightness temperature of the snow pack (eg. Brucker *et al.*, 2011). The spectral gradient of the brightness temperature is the basis for the workhorse algorithms of the retrieval of snow accumulation from remote sensing (eg. Hallikainen and Jolma, 1986; Chang *et al.*, 1997). These differential scatter darkening algorithms must be tuned separately for each region (eg. Tait, 1998). Tedesco *et al.* (2010) recommend dynamic tuning coefficients based on snow evolution models driven by weather data, and, since an open-loop model isn't sufficient by itself, with ground truth measurements.

These in-situ sensors record these important snow pack point parameters for the duration of the snow season with user-selectable temporal resolution as fine as 1 minute. They consume about 0.4 mW in their quiescent state, and about 100 mJ for every measurement of all the parameters, which takes about 3 seconds to complete. Thus, measuring once every 5 minutes, it consumes less than 1 mW on average, which prior experimentation has determined as an upper limit to power dissipation to prevent observable metamorphism due to the heat created.

The sensors are configured via a serial interface that plugs into a USB port. Configuration is performed by the user with simple typed commands in standard interfaces, such as Arduino's SerialMonitor which works in Linux, Mac and Windows environments. As a data logger, the data is not available until the unit is retrieved from the environment and the data downloaded via the serial interface. In the works is a wireless interface. The hardware supports such an interface, but the software must be written to do this.

DESCRIPTION OF THE TECHNOLOGY

Figure 1 depicts two of the in-situ snow sensor modules. The unit on the left has its sensor face towards the camera. The white printed circuit board (PCB) contains the sensor for dry snow density and snow moisture. A microstrip resonant circuit is under the white solder mask, and electric fields traveling on this circuit penetrate the space above the PCB, and so the material in contact with the PCB will affect the circuit performance. To the right of the PCB in the unit on the left in Fig. 1, and in the right in Fig. 2 (bottom), is an optical link for grain size characterization. A byproduct of this measurement is a quantification of the ambient light level, 'tho this measurement has not been calibrated yet. Finally, electronic thermometers on the main (green) PCB measure the ambient temperature of the snow pack.

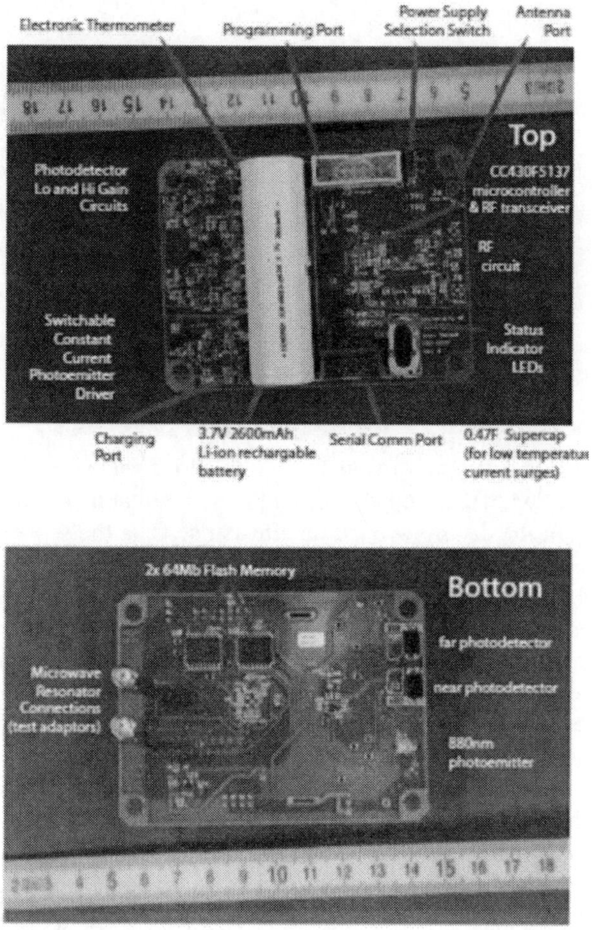

Figure 2. Electronic circuits of the main printed circuit board (PCB).

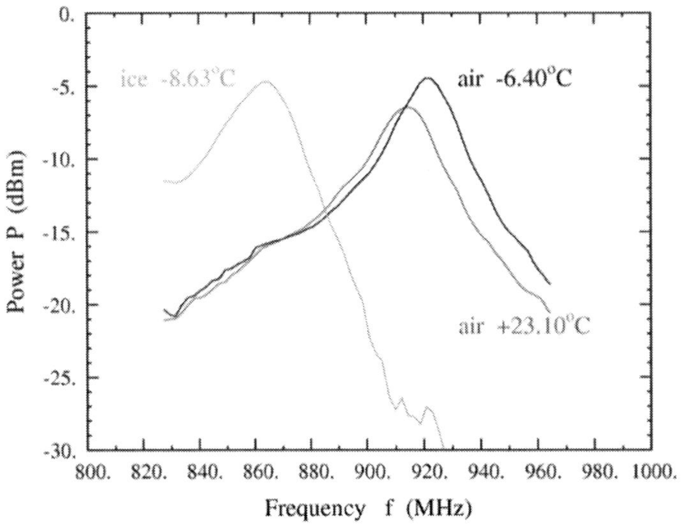

Figure 3. Laboratory measurements of the microwave resonator responses of unit 41082B for air and solid tap water ice, at the temperatures indicated.

The microwave circuit is set up as a resonator. Continuous wave (ie. single frequency) power is swept from 827 MHz to 963 MHz in 64 steps, and the power that makes it through the resonator is measured. The density of the dry snow pack manifests itself via the snow pack dielectric constant as a reduction in the resonant frequency of the resonator. Moisture also affects the resonator, but in addition to moving the resonant frequency, moisture also reduces the power through the device, and manifests itself as an increased resonant bandwidth. To a much lesser degree, temperature also changes the resonant performance. Figure 3 shows some calibration data for unit 41082B, in which two measurements of air, at room temperature and below freezing, are compared to each other and to solid ice. The resonant frequency is inversely related to the microwave index of refraction, and, for dry snow, this is directly related to the snow density.

The optical link is used to characterize the size of the snow grains. An infrared emitter sends pulses of infrared light with a peak wavelength of 880nm through the clear polycarbonate case into the snow pack. To compensate for ambient light level within the snow pack, the emitter is toggled on and off at a rate of 1 kHz. Two infrared detectors, one "near" at 2 cm from the emitter and one "far" at 3 cm from the emitter, measure the changes in light level due to scattering in the snow pack. Active filters remove the DC level, due to ambient light, and the detector output is recorded at 2kHz, with alternate data of the emitter on and off states. The difference of these states is the useful data from the individual detectors. The further difference of the near data minus far data constitutes a double difference, and this double difference is strongly dependent on the grain size. Figure 4 shows this double difference for halite (table salt) that has been ground and sieved into different crystal sizes. The double difference is roughly proportional to the square root of the mean grain diameter, for grain diameters from 0.1 mm to 2.5 mm. These measurements of salt are performed for each unit, and the data regressed as a power law relationship. The distinction between the halite crystals and distilled water ice chips is less than the error in the measurements of the ice, so we do not (yet) make an adjustment for the fact that the media have different optical indices of refraction.

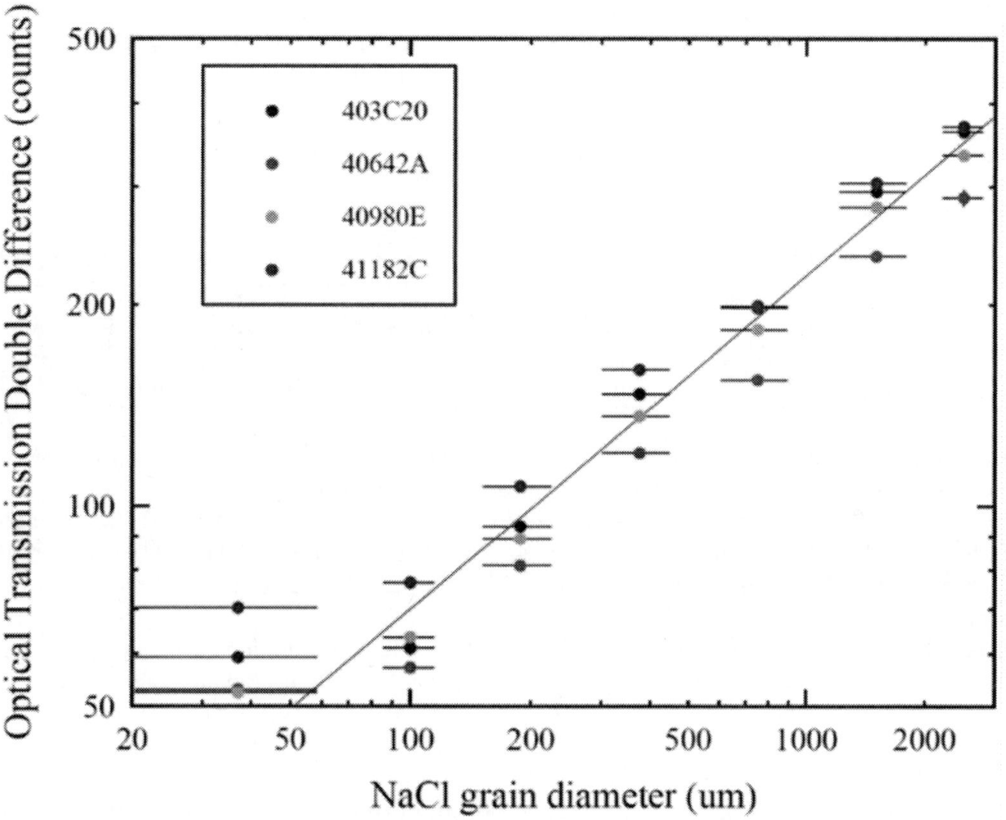

Figure 4. Measurements of the optical double difference vs. halite grain size, for different units. The halite serves as a stand-in for ice.

DATA FROM THE FIRST SEASON

Figures 5-7 show what this hardware can do. These are time-series of hourly snow pack measurements, for the winter of 2016-2017 at the Mesa Lakes Ranger Station on Grand Mesa, CO, USA. The Ranger Station was part of the Local Scale Observation Site (LSOS) of NASA's Snow Experiment (SnowEx) year 1 campaign. Snow sensor units were deployed in pairs on the days in Table I.

The first pair were deployed in October, well before snow began to accumulate. Thus, these units gathered data at the base of the snow pack, when it did start to accumulate on 2016 Nov 18 (day of water year 48) as provided by the Mesa Lakes SNOTEL (Schaefer and Paetzold, 2000; http://wcc.sc.egov.usda.gov/nwcc/site?sitenum=622). The last two pairs were deployed during the Intensive Observation Period (IOP) of SnowEx, and were retrieved at the end of the IOP for initial data evaluation. As such, these units only collected a few days of data. Three pairs were deployed in the middle of December, when a major snow storm added significant snow to the pack. The SNOTEL reports a peak snow depth of 168cm on Jan 25 (dowy 116) and peak snow water equivalent (SWE) of 480mm in early April (dowy 188-193). The last day of measurable snow at the SNOTEL was Jun 01 (dowy 243).

Table 1. Snow sensor deployment parameters for SnowEx year 1 campaign. The date of last data represents the day in which the snow sensor was retrieved, except for units 405C3C, 403415, and 41142B (in red), for which data collection stopped when the battery was exhausted.

Snow Sensor Deployment Data					
Snow depth at time of deployment	Date of first data	Snow Sensor 1 serial number	Date of last data, SS1	Snow Sensor 2 serial number	Date of last data SS2
0 cm	2016/10/02	40DC14	2017/05/30	405C3C	2016/12/08
47 cm	2016/12/15	40F404	2017/04/21	403415	2017/03/29
82 cm	2016/12/16	40142B	2017/04/21	41342D	2017/04/21
125 cm	2016/12/17	40B82F	2017/04/21	41142B	2017/02/05
134 cm	2017/02/20	40642A	2017/02/25	41082B	2017/02/25
134 cm	2017/02/22	40E82A	2017/02/25	41182C	2017/02/25

Deployment consisted of placing the in-situ sensor on top of the existing snow pack (or the ground) with a long wall of the unit down (ie. same orientation as in Fig. 1). The sensor face was vertical and the entire unit oriented so that the sensor face was to the north. The pairs were placed somewhat at random, but approximately 1m apart, in a 3m x 3m portion of the LSOS reserved for SnowEx ground truth measurements. This portion of the LSOS was not used by other SnowEx groups for any purpose.

All but three of the in-situ snow sensors collected data for the duration of their deployment. These three stopped collecting data due to loss of charge on their lithium ion batteries. It was later determined that these units were subjected to a deep discharge that likely damaged the batteries during initial testing of the unit's hardware and software in the laboratory. These batteries (and those for a few others) will be replaced before these units are deployed again. The evolution of the software since these deep discharge events has significantly lowered power consumption, reducing the risk of these events, and the expected implementation of a shut-down routine automatically triggered by low voltages will further minimize the likelihood of these damaging event.

For clarity in the figures to follow, data from only six of these 12 units will be shown: those with the longest time series that coincides with the presence of the snow pack.

Figure 5 shows the temperature record retrieved from six of these units. The temperatures oscillate wildly for those that are not yet buried by snow, as they experience the daily air temperature swings. Around dowy 90, shortly after the major snow storms of December, the temperatures rise, and the oscillations dampen, as one goes deeper into the snow pack. The snow pack warms as the winter season progresses, until around dowy 165, when all of the thermometers flatline. This indicates that the entire snow pack is melting. On dowy 177, unit 403415 suddenly gives readings of much colder temperatures. Shortly after this event, this unit ceased logging data, indicating a discharged battery: the low temperature readings are anomalous side effect of its low power. On dowy 203 most of the snow sensors were retrieved from the still-melting snow pack so that results could be reported at the Eastern Snow Conference.

There is a significant negative bias of about one degree Celsius in these temperature readings. The cause is currently under investigation. However, the isothermal behavior of the snow pack in early Spring makes it obvious what the bias is. Indeed, one can see that the temperatures at the same height in the snow pack are not merely highly correlated, but are in near agreement, throughout the winter. Thus, should we need an empirical correction, it is easy to obtain.

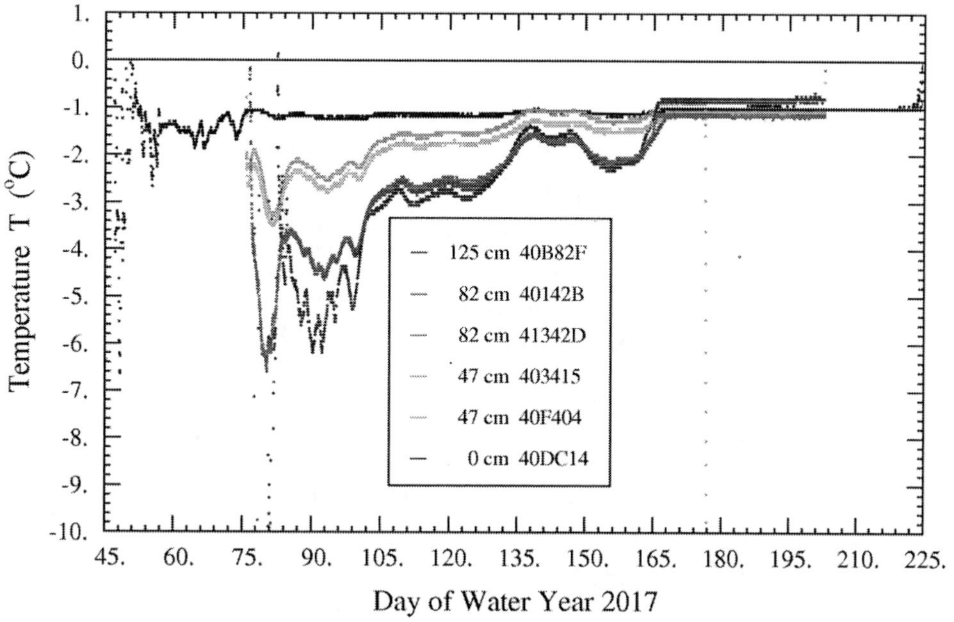

Figure 5. Thermometer readings of in-situ snow sensors embedded in a snow pack at the Ranger Station on Grand Mesa CO during SnowEx year 1. The legend indicates the depth of the snow pack at the time the snow sensor was deployed, which are followed by the instrument's serial number.

We are currently investigating whether the source of the temperature bias is a reference voltage used for the analog-to-digital converter that is distinctly different near freezing than it is at room temperature, or if the digital output voltage level used to power the electronic thermometer is insufficient for the thermometer's specified accuracy. The former would require another calibration datum, while the latter may require the main printed circuit board to be revised to have the digital output switch the battery voltage to drive the thermometer rather than directly powering it.

Figure 6 shows the partly calibrated snow density data retrieved from the in-situ snow sensors. The shift of the resonant frequency from that of air is used to find a dielectric constant. This dielectric constant is then used to infer the snow bulk density, but the possible influence of moisture on the dielectric is not considered in this figure. Thus, the figure shows sudden and dramatic increases in reported snow density at the onset of melt on dowy 165. The measured resonator quality factor will be used for the detection and quantification of moisture. The black curve, for unit 40DC14 at the bottom of the snow pack, uses an earlier version of the resonator which is noisier than the other units that are equipped with the latest resonator. It also shows a more enhanced response to moisture, which can be most clearly seen in the relative warm period from dowy 135 thru 150. The effect of the major snowstorm of dowy 75-77 on the snow density can be clearly seen: the density at the bottom of the snow pack climbs from 150 kg/m^3 to 250 kg/m^3, and the sensors further up in the snow pack show sudden rises that correspond to them getting buried. The two snow sensors at 47cm depth show very good agreement with each other before slowly diverging as the winter progresses. The snow that fell in the middle of the storm (monitored by the sensor at 82cm) was warmer and heavier than the snow both earlier and later in the storm, as evidenced by its greater initial density. The snow that fell at the end of the storm was much colder, and was much lighter, and the snow sensor 40B82F could not be deployed directly on the surface of the snow pack, but rather sank about 15cm into the snow pack to where the snow was sufficiently dense to support it. So while it is appropriate that its readings are of a

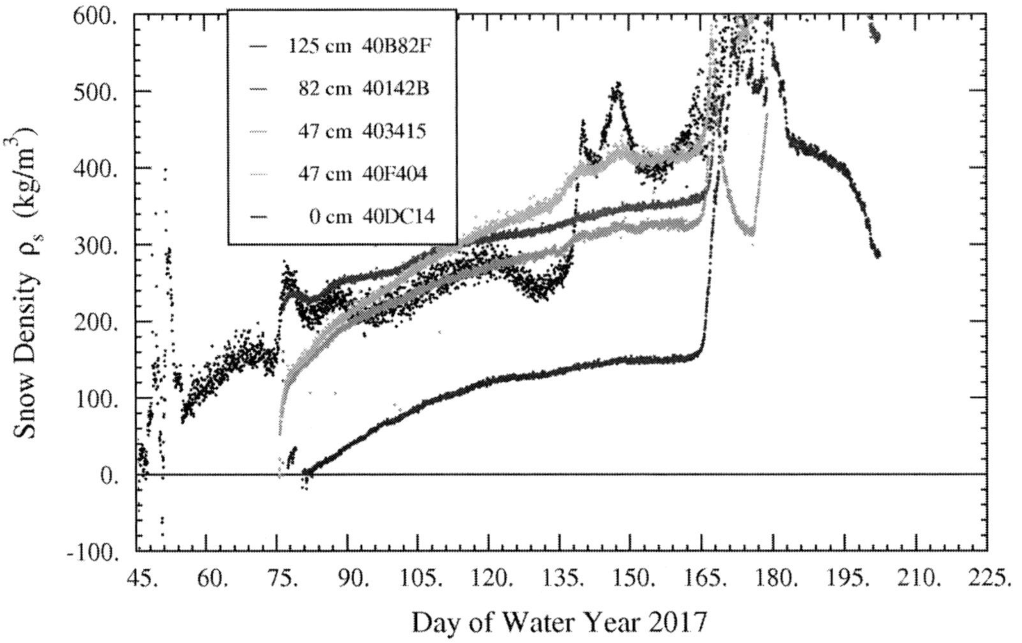

Figure 6. Retrieved snow density from snow sensors embedded in a snow pack at the Ranger Station on Grand Mesa CO during SnowEx year 1. The legend is the same as for the previous figure. The density is derived from the dielectric constant, assuming a dry snow pack. Effects of moisture, most obvious after the snow started melting around dowy 165, have not been removed. Unit 40DC14 (black curve) uses an earlier version of the resonator board which is both noisier and more sensitive to moisture than the other units.

less dense snow pack, we suspect that an air gap between the resonator and the snowpack developed that was slowly filled in over time. All of these sensors show a general trend of increasing density over time as compaction occurs.

Figure 7 shows the partly calibrated optical diameter of snow grains retrieved from the snow pack. The solid lines are for snow sensor units that have been individually calibrated with up to seven distinct sizes of halite crystals. The two dashed curves are not individually calibrated, and so their readings are suspiciously high, but they shown for their temporal trends. (Unit 40B82F is on the UMich truck, in Colorado at the time of this writing, and unit 41342D had a fabrication fault that renders its calibration suspect.) For the most part, the grain sizes grow slowly over time, as expected. Two notable exceptions are 40B82F (blue curve) and 40DC14 (black curve). Unit 40B82F shows a decrease in grain size from dowy 84 to 105, but this is likely an artifact of the issues associated with the deployment, as discussed above, until the snow full packed around the unit. The sudden jumps in the data prior to dowy 84 bolster this interpretation. The snow sensor at the bottom of the snow pack exhibits very high values of its light link level, and oscillates with an almost monthly period. This is interpreted as the development of depth hoar, and possibly growth of ice crystals directly on the in-situ snow sensor package. The onset of melt at dowy 165 results in sudden change in the character of the time series, and should be considered the end of the valid data.

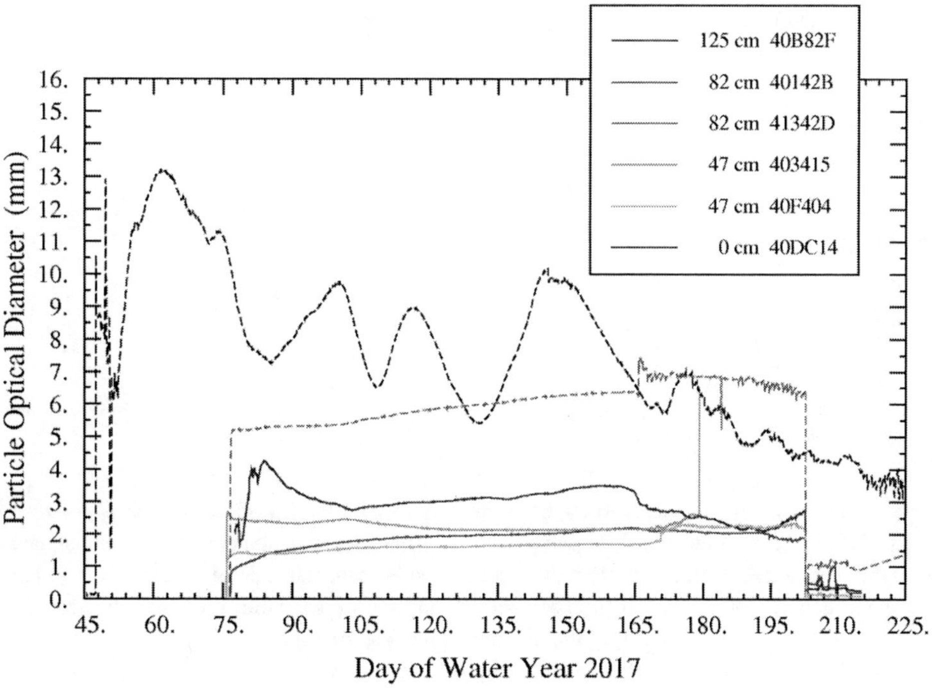

Figure 7. Retrieved snow grain optical diameter. The dashed curves are for units for which the calibration coefficients are an average of those units that have been individually calibrated, and should be considered for their trends, not their magnitude.

While the separate in-situ sensors show considerable consistency with each other, independent measurements are needed to assess their accuracy. This is why the SnowEx LSOS was chosen for the deployment of these units. In the morning of 2017 Feb 14 (dowy 136), two snow pits were simultaneously dug at the Ranger Station, and a standard suite of manual measurements were undertaken by two snow research groups not affiliated with the University of Michigan. The results of those manual measurements are shown in Fig. 8. The reported snow depth at the time was 150cm. Unfortunately, these are the only two reports of snow pits dug at the Ranger Station, but what is heartening is that the reports agree well with each other.

The manual temperature profiles agree quite well with that recreated from the in-situ instruments, and the agreement is likely better than that shown if the temperature bias identified above is removed.

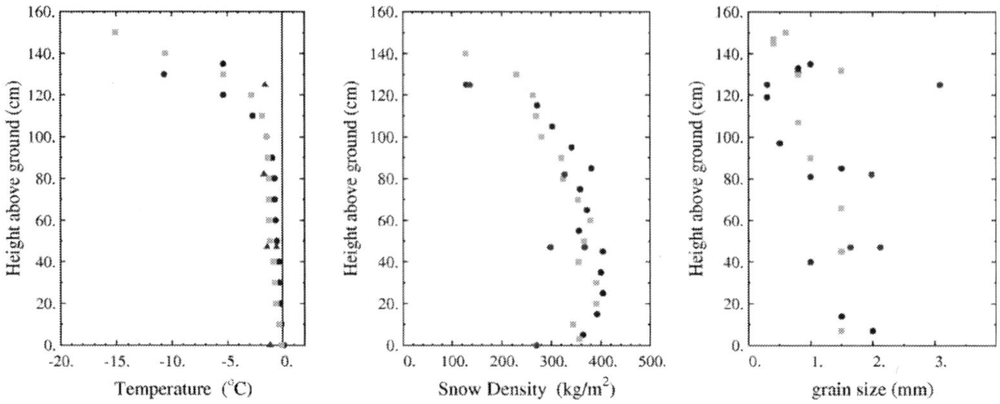

Figure 8. Comparison of data acquired by the snow sensors at 08:00 MST on 2017 Feb 14 (dowy 136) with two sets of manual measurements at the same location. Blue and cyan marks indicate the manual snow pit data, while the magenta points are from the electronic snow sensors. The snow depth was 150cm. The uppermost snow sensor was not in good contact with the snow pack, while the lowest utilized an earlier version of the instrument that measured density.
Left: temperature profile; middle: snow density profile; right: grain size profile.

CONCLUSIONS

The University of Michigan has developed a small, self-contained datalogger that is intended for operation within a snow pack, in order to measure those point properties of snow packs which typically evolve over a snow season. These properties include temperature, density, wetness, grain size and ambient light. Twelve of these were deployed for varying lengths of time at the Mesa Lakes Ranger Station in Grand Mesa National Forest, Colorado, during the year 1 campaign of the NASA Snow Experiment (SnowEx). These instruments collected data at the top of every hour. While these instruments were developed with a focus on low cost and time-series format of data collection, the data, calibrated in the lab with proxy materials only, compare favorably with limited manual stratigraphy measurements at the site. The temperatures appear to have a cold bias of about 1C. The snow density measurements agree with manual measurements. The optical grain sizes have a high bias, but with several of the instruments experiencing technical issues related to different manufacturing procedures, it is difficult to quantify the magnitude of the bias at this time. The ambient light and wetness measurements have not yet been calibrated.

After several different electrical and mechanical configurations of these instruments have been attempted, this data set points to the best configuration of the electronic sensors. We are modifying the units that had alternative configurations to this best configuration, and currently have 15 units in this configuration and expect to have a total of 23 units ready for the winter 2017-18 field season.

REFERENCES

Brucker L, Picard G, Arnaud L, Barnola J-M, Schneebeli M, Brunjail H, Fily M. 2011. Modeling time series of microwave brightness temperature at Dome C, Antarctica, using vertically resolved snow temperature and microstructure measurements. *Journal of Glaciology*, **57**(201): 171-182.

Chang ATC, Foster JL, Hall DK, Goodison BE, Walker AE, Metcalfe JR, Harby A. 1997. Snow parameters derived from microwave measurements during the BOREAS winter field campaign. *Journal of Geophysical Research*, **102**(D24): 29,663-29,671.

Hallikainen MT, Jolma PA. 1986. Retrieval of the water equivalent of snow cover in Finland by satellite microwave radiometry. *IEEE Transactions on Geoscience and Remote Sensing*, **24**(6): 855-862.

Schaefer GL, Paetzold RF. 2000. SNOTEL (SNOwpack TELemetry) and SCAN (Soil Climate Analysis Network). In *Proceedings*, Automated Weather Stations for Applications in Agriculture and Water Resources Management: Current Use and Future Perspectives, Lincoln, NE.

Tait AB. 1998. Estimation of snow water equivalent using passive microwave data. *Remote Sensing of Environment*, **64**(3): 286-291.

Tedesco M, Reichle R, Low A, Markus T, Foster JL. 2010. Dynamic approaches for snow depth retrieval from spaceborne microwave brightness temperature. *IEEE Transactions on Geoscience and Remote Sensing*, **48**(4): 1955-1967.

Oral Session #3: Remote Sensing of SWE and Snow Depth

Validation of Physical Model and Dual-Frequency Radar Retrieval Algorithm using SnowSAR Data

JIYUE ZHU[1], SHURUN TAN[1], JOSHUA KING[2], CHRIS DERKSEN[2], CHUAN XIONG[3], JUHA LEMMETYINEN[4] AND LEUNG TSANG[1]

ABSTRACT

In this paper we combine the bicontinuous dense media radiative transfer (Bic-DMRT) model with a radar retrieval algorithm of snow water equivalent (SWE). Both the Bic-DMRT model and the retrieval algorithm are validated using airborne SnowSAR data at X and Ku band. The retrieval algorithm is based on the absorption loss of the snowpack which is directly proportional to the SWE. In the algorithm, Bic-DMRT is first applied to generate the look-up table (LUT) of scattering properties of snow at X- and Ku-band. With the LUT, we plot the scatter diagram of the scattering albedo $\boldsymbol{\omega_{Ku}}$ **VS.** $\boldsymbol{\omega_X}$ and the optical thicknesses $\boldsymbol{\tau_{Ku}}$ **VS.** $\boldsymbol{\tau_X}$, respectively and derive regression formulas for $\boldsymbol{\omega_{Ku}}$ **VS.** $\boldsymbol{\omega_X}$ and $\boldsymbol{\tau_{Ku}}$ **VS.** $\boldsymbol{\tau_X}$ separately. Furthermore, regression formulas between multiple (volumetric) and single scattering at both bands are also derived. With these four regression formulas, we transform the volumetric scattering of a single snow layer at dual bands as functions of $\boldsymbol{\omega_X}$ and $\boldsymbol{\tau_X}$. The background scattering is then subtracted from the total scattering which gives the volume scattering of snow. Based on the obtained volume scattering, a cost function is established to find the solution. Performance of the retrieval algorithm was tested using SnowSAR data (the Finland SnowSAR1 and SnowSAR2, and the Canada SnowSAR campaigns). The retrieval algorithm achieves RSME smaller than 20 mm for all the SnowSAR data in this paper which meets the requirements for snow mass mission concept studies currently underway at the European Space Agency (ESA) and Canadian Space Agency (CSA).

Keywords: bicontinuous dense media radiative transfer (Bic-DMRT), retrieval algorithm, snow water equivalent (SWE), look-up table (LUT), regression formula, SnowSAR

ESTIMATION OF BACKGROUND SCATTERING

The radar observation consists of two parts, the volume scattering component from the snowpack and the background scattering component from the snow-free ground, as shown in Figure 1 (a). These two components are combined to give the expression of radar observation below:

[1] Jiyue Zhu, Shurun Tan and Leung Tsang are with Radiation Laboratory, Department of Electrical Engineering and Computer Science, The University of Michigan, Ann Arbor, 48109-2122 MI USA.

[2] Joshua King and Chris Derksen are with Climate Research Division, Environment and Climate Change Canada, Toronto, ON M3H 5T4, Canada

[3] Chuan Xiong is with Institute of Remote Sensing and Digital Earth, Chinese Academy of Sciences, Beijing, China

[4] Juha Lemmetyinen is with Arctic Research Centre, Finnish Meteorological Institute P.O.Box 503, Fin-00101 Helsinki Finland

$$\sigma_{pq}^{tol} = \sigma_{pq}^{g} \exp\left(-\frac{2\tau}{\cos\theta_t}\right) + \sigma_{pq}^{snow} \quad (1)$$

where σ_{pq}^{snow} is the volume scattering component from the snowpack and σ_{pq}^{g} is the scattering from snow-free ground which is attenuated by the snow layer by a factor of $exp(-2\tau/cos\theta_t)$. τ is the optical thickness of the snow layer and θ is the transmitted angle in the snow. Figure 1 (b) illustrates the data of SnowSAR measurements from the flights in Canada and the bicontinuous / DMRT LUT. The range for Canada SnowSAR raw data are from -18dB to -11dB and from -11dB to -6dB, respectively. The volume scattering component is obtained by subtracting the background scattering from Canada SnowSAR raw data with equation **Error! Reference source not found.**. The range of volume scattering component of Canada SnowSAR is from -22dB to -14dB for X-band and from -13dB to -7dB for Ku-band. The shift between volume scattering component and raw data at X-band is more than that at Ku-band which indicates that background scattering component from the ground has more impacts on X-band. From Figure 1 (b), we find all of volume scattering component from Canada SnowSAR are located within the range that bicontinuous / DMRT model predicts which means variabilities exhibited in Canada SnowSAR data are covered by the bicontinuous / DMRT LUT. The bicontinuous / DMRT works well in modeling the volume scattering for Canada SnowSAR data.

The surface backscattering at co- and cross-pol taken from a prebuilt lookup table based on NMM3D (Huang, 2010 and 2012) will be used in further studies, however, the table does not cover the full surface roughness range in our campaigns. Right now we apply an empirical model proposed by (Oh, 1992) to calculate the scattering from bare soil surfaces.

PARAMETERIZATION AND REGRESSION TRAINING

With the snowpack parameters, we utilize the bicontinuous / DMRT model to generate a look-up table (LUT) (Tan, 2015). We tabulate 6 bicontinuous / DMRT outputs, including four parameters which are the scattering albedo ω_X, ω_{Ku} and the optical thicknesses τ_X, τ_{Ku} at X and Ku band and two observations $\sigma_{VV}^{X}, \sigma_{VV}^{Ku}$ which are the VV polarization backscatter of snowpack at dual bands. With the LUT, we derive regression formulas from the scattering albedo ω_{Ku} vs. ω_X and the optical thicknesses τ_{Ku} vs. τ_X separately, as shown in Figure 2. Next, given the single scattering solution which only depends on ω and τ, the regression formulas between single and multiple scattering are derived to transform σ_{VV}^{X} as a function of ω_X, τ_X and σ_{VV}^{Ku} as a function of ω_{Ku}, τ_{Ku}, , as illustrated in Figure 3. We then reduce the number of unknowns by utilizing the regression formulas of ω_{Ku} vs. ω_X and τ_{Ku} vs. τ_X. Finally, both $\sigma_{VV}^{X}, \sigma_{VV}^{Ku}$ only depend on ω_X, τ_X resulting in 2 observations and 2 unknowns (Cui, 2016).

$$\sigma_{VV}^{X}\ (\text{dB}) = -3.3841481 + 0.92035737 \times 10\log_{10}\left\{0.75\cos\theta_t \omega_X \left[1-\exp\left(-2\tau_X/\cos\theta_t\right)\right]\right\}$$

$$\sigma_{VV}^{Ku}\ (\text{dB}) = -0.24752984 + 1.1091572 \times 10\log_{10}\left\{\begin{array}{l} 0.75\cos\theta_t\left(\dfrac{\omega_X}{0.58352316\omega_X + 0.43973192}\right)\times \\ \left[1-\exp\left(-2\exp\left(1.3760377 + 0.96250819\ln\left(-\tau_X\right)\right)/\cos\theta_t\right)\right]\end{array}\right\} \quad (1)$$

Originally there are four snowpack parameters in the bicontinuous / DMRT model. After parameterization, we reduce the number of unknowns to two. The parameterized model only depends on ω_X, τ_X. Furthermore, the parameterized model provides with an effective equivalent single layer description for the multi-layer snow. With such characterization of snowpack, we directly retrieve ω_X and τ_X in the retrieval algorithm. The retrieved ω_X and τ_X give the absorption loss of the snowpack, which is proportional to SWE. In addition, it is also easy to obtain the priori information for the retrieval by using the parameterized model.

VALIDATION OF THE PARAMETERIZED MODEL AND PERFORMANCE OF THE RADAR RETRIEVAL ALGORITHM USING EXISTING SNOWSAR DATA

Three SnowSAR datasets have been used for validation of the algorithm:

a) SnowSAR1 airborne campaign has been carried out in the Lappish region of Finland from March 12th to 19th, 2011 and mostly focused on the area close to the town of Sodankylä. In the CoReH2O, a dual frequency, X- and Ku-bands, dual polarization (VV&HV) mini-SAR airborne system (SnowSAR) has been deployed (CoReH2O, 2010).

b) SnowSAR2 airborne campaign is the continuous campaign of SnowSAR1, which was also performed around Sodankyla, Finland, from December 19th, 2011 to March 24th, 2012 (Cohen, 2015).

c) The Canadian SnowSAR campaign is another airborne campaign with SnowSAR instrument performed within the Track Valley Creek (TVC), the Northwest Territories, Canada, during winter 2013. Two sets of polarimetric SAR images operating at X- and Ku-bands were obtained (King, 2017).

Figure 4 depicts that the results of parameterized model agrees well with the Canada SnowSAR data at both X- and Ku- band. The root-mean-square error (RMSE) between model and Canada SnowSAR data is ~0.3dB and their correlation coefficient is above 0.94. It indicates that the model predicts the backscattering of snowpack accurately provided reasonable physical parameters of the snowpack.

We apply a statistical method to search for the solution. In practice, a priori constrained least-squares algorithm is used to fit the model predictions into radar observations at dual bands, which was also the case in the CoReH2O SWE retrieval (Cui, 2016 and CoReH2O, 2010). The cost function is defined as:

$$F = \underset{\omega_X,\tau_X}{MIN}\left\{\frac{\left(\sigma_{VV}^{X,obs}-\sigma_{VV}^{X,\mathrm{model}}(\omega_X,\tau_X)\right)^2}{2s_X^2}+\frac{\left(\sigma_{VV}^{Ku,obs}-\sigma_{VV}^{Ku,\mathrm{model}}(\omega_X,\tau_X)\right)^2}{2s_{Ku}^2}+\frac{\left(\omega_X-\overline{\omega_X}\right)^2}{2s_{\omega_X}^2}\right\} \tag{2}$$

where $\sigma_X^{obs}, \sigma_{Ku}^{obs}$ are the radar observations at dual bands including surface scattering from the bottom ground and volume scattering from the snowpack, $\sigma_{VV}^{X,model}, \sigma_{VV}^{Ku,model}$ are the backscattering predictions from the regression forward model, $\overline{\omega}_X$ is the a priori value for albedo. s_X^2, s_{Ku}^2 are the expected error standard deviation of the measurements which is also regarded as the variance of speckle in a SAR image, and $s_{\omega_X}^2$ act as the variance of the priori constraint. The contribution from each term in cost function is normalized by assuming a Gaussian distribution. ω_X, τ_X are the two parameters we need to solve from this equation. With cost function, we find the best fit ω_X, τ_X for each set of $\sigma_X^{obs}, \sigma_{Ku}^{obs}$.

We have validated the radar retrieval algorithm by applying the SnowSAR data. We set $s_X = 0.5, s_{Ku} = 0.6$ (Cui, 2016). Based on the error standard deviation of the measurements which is also regarded as the variance of speckle in a SAR image. And we set the variance of the priori constraint as $s_{\omega_X} = 0.1, s_{\tau_X} = 0.01$ based on the variance of ω_X, τ_X. The priori parameter ω_X for each data points of SnowSAR was taken from its corresponding matched dataset in LUT in the forward model validation. τ_X as uniformly sampled in a range which is determined by datasets who have the same ω_X. The retrieval results are illustrated in Figure 5, Figure 6, and Figure 7, for the Finland SnowSAR1, the Finland SnowSAR2, and the Canada SnowSAR, respectively. These results meet the requirement defined in SCLP (the NASA Snow and Cold Land process). SCLP requires that the RMSE of SWE retrieval be less than 20mm for shallow snowpack with SWE less than 200mm and the RMSE be less than 10% of the total SWE when the SWE is larger than 200mm.

In Figure 5 of the results of the Finland SnowSAR1 dataset, both the RMSE and the bias are small. The relatively large correlation coefficient indicates that the retrieval results reflect the variation of the measurements. The fact that the range of the retrieval results is slightly larger than that of the observed SWE is possibly due to the inaccurate guess of the scattering from soil and vegetation. Meanwhile, the correlation in Figure 6 for the Finland SnowSAR2, and the bias and correlation in Figure 7 for the Canada SnowSAR datasets are better than that of SnowSAR1 in Figure 5. In Figure 6, and Figure 7, the correlation coefficient is approaching 1 and the ratio of retrieval and measurement data is almost 1:1, indicating excellent retrieval performance. The reason is that for SnowSAR2 and Cananda SnowSAR data we have selected data points located in open areas and the guess of the ground scattering is much better which improves the performance of the algorithm. The achieved RMSE and the correlation coefficient are 15.22 mm and 0.739 for the Finland SnowSAR1dataset, 19.07mm and 0.782 for the Finland SnowSAR2 dataset, and 15.89mm and 0.916 for the Canada SnowSAR dataset, respectively.

.

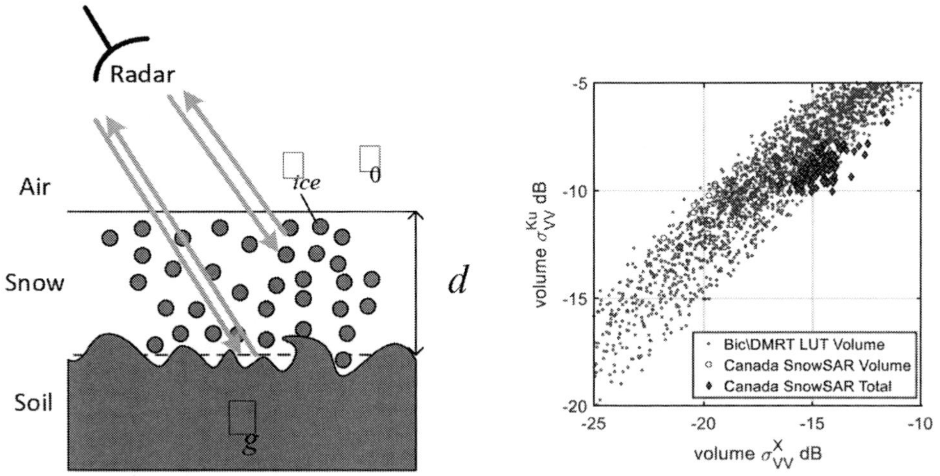

Figure 1. (a, left) Illustration of volume scatteringa and surface scattering; (b, right) surface scattering affects the total scatter more in X- than in Ku- band.

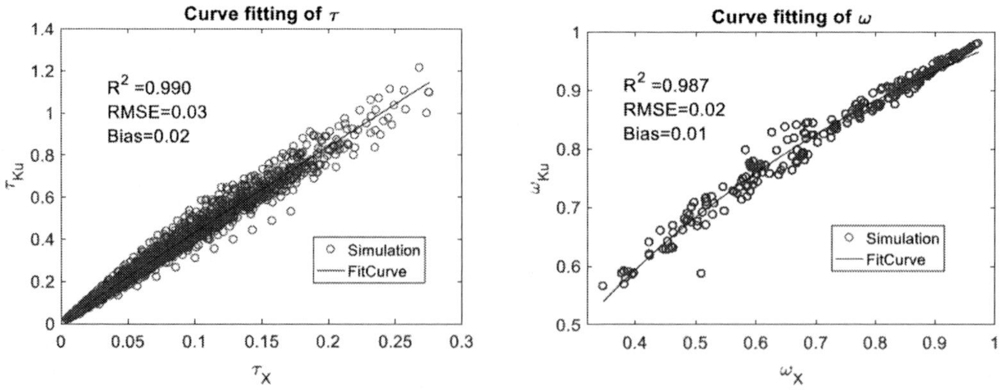

Figure 2. Comparison of scattering parameters between simulation and regression results: optical thickness (a, left) and scattering albedos (b, right).

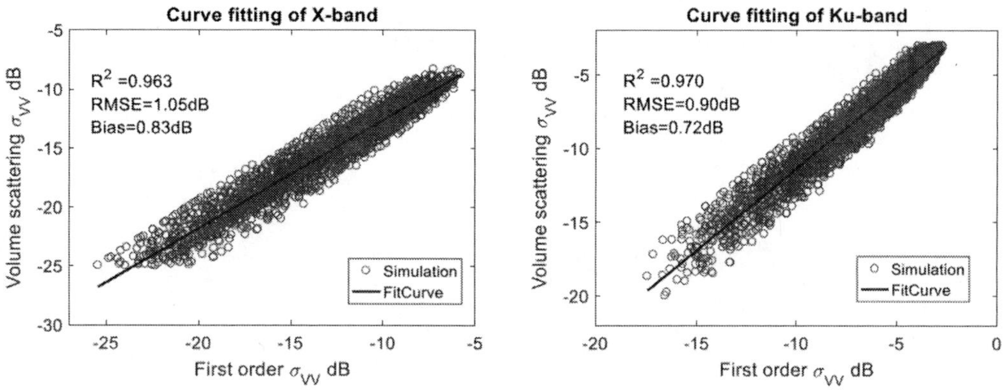

Figure 3. Comparison of volume scattering and first order regression at X-band (a, left) and Ku-band (b, right).

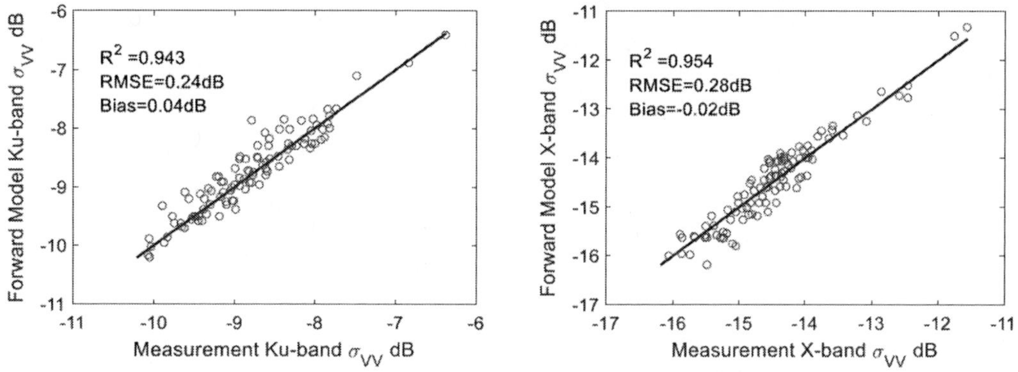

Figure 4. Validation of parameterized model Canada SnowSAR data at X-band (a, left) and Ku-band (b, right).

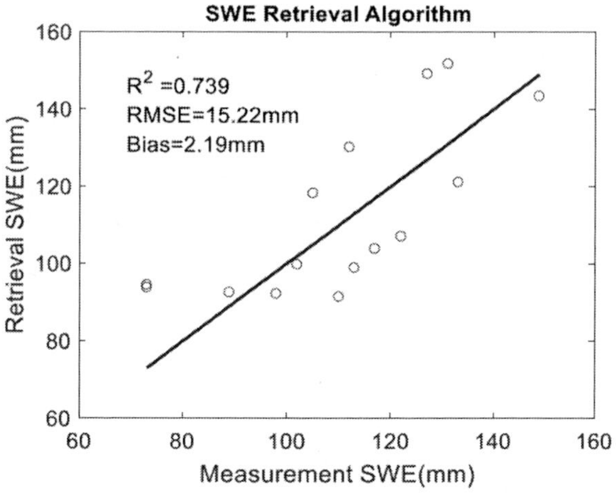

Figure 5. Comparison of SWE retrieved with SWE measured for SnowSAR1 data.

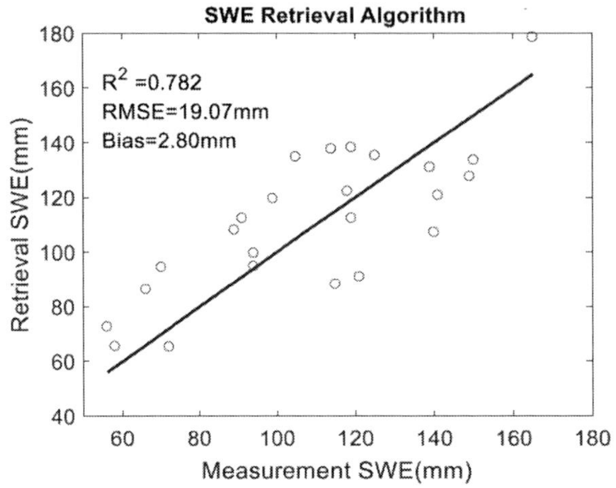

Figure 6. Comparison of SWE retrieved with SWE measured for SnowSAR2 data.

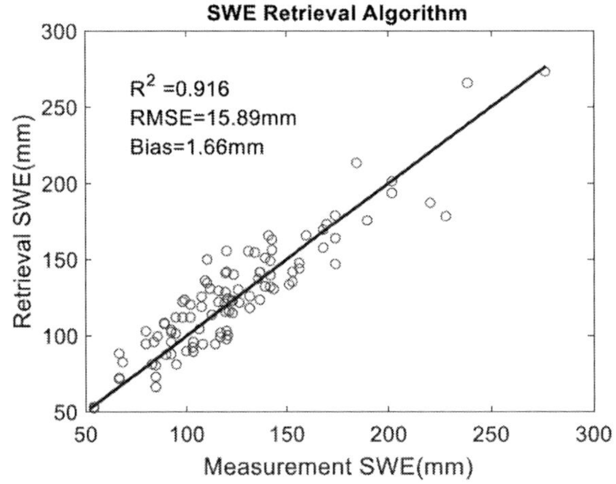

Figure 7. Comparison of SWE retrieved with SWE measured for Canada SnowSAR data.

REFERENCES:

CoReH2O. 2010. Technical assistance for the development and deployment of an X- and Ku- band miniSAR airborne system. Doc. No. MS-Snowsar-D7, issue 2.0.

Cohen J *et al.* 2015. The effect of boreal forest canopy in satellite snow mapping—a multisensor analysis. *IEEE Transactions on Geoscience and Remote Sensing*, **53**: 6593-6607.

Cui Y *et al.* 2016a. Estimating Snow Water Equivalent with Backscattering at X and Ku Band Based on Absorption Loss. *Remote Sensing*, **8**: 505.

Huang S. 2010. Backscattering coefficients, coherent reflectivities, and emissivities of randomly rough soil surfaces at L-band for SMAP applications based on numerical solutions of Maxwell equations in three-dimensional simulations. *IEEE Transactions on Geoscience and Remote Sensing*, **48**: 2557–2568.

Huang S, Tsang L. 2012. Electromagnetic scattering of randomly rough soil surfaces based on numerical solutions of Maxwell solutions of Maxwell equations in three-dimensional simulations using a hybrid UV/PBTG/SMCG method. *IEEE Transactions on Geoscience and Remote Sensing*, **50**: 4025–4035.

King J *et al.* 2017. Exploring the influence of snow microstructure on dual-frequency radar measurements. *2017 IEEE Geoscience and Remote Sensing Symposium.*, Fort Worth, USA, Jul. 23-28 2017.

Oh Y *et al.* 1992. An empirical model and an inversion technique for radar scattering from bare soil surfaces. *IEEE Transactions on Geoscience and Remote Sensing*, **30**: 370-381.

Tan S *et al.* 2015. Modeling both Active and Passive Microwave Remote Sensing of Snow Using Dense Media Radiative Transfer (DMRT) Theory with Multiple Scattering and Backscattering Enhancement. *IEEE Journal of Selected Topics in Applied Earth Observations and Remote Sensing*, **8**: 4418-4430.

Oral Session #6: Snow Modelling and Snow Cover

Cloud detection in mixed-snow-vegetation areas using machine learning

NAN CHEN,[1] WEI LI,[1] TOMONORI TANIKAWA[2], MASAHIRO HORI[3], RIGEN SHIMADA[3], KNUT STAMNES[1]

EXTENDED ABSTRACT

Keywords: cloud detection, snow, vegetation, machine learning

Cloud detection is a critically important first step required to derive many satellite data products. Chen *et al.* (2014) presented a novel reflectance-based cloud detection algorithm (SCM) designed for snow/ice-covered areas. Validations using collocated CALIOP and MODIS Aqua observations over Greenland showed superior performance of the SCM compared to the MODIS Cloud Mask. Now we are extending our work to mid-latitude areas, where snow is frequently mixed with forest and soil, by using advanced machine learning techniques in conjunction with detailed radiative transfer modeling. Statistical results obtained by using collocated CALIOP and MODIS data show consistent and superior performance of our SCM algorithm compared to the MODIS Cloud Mask during the winter seasons over mid-latitude snow-covered areas.

A cloud mask is always the first priority task for satellite remote sensing over land, ocean, or cryosphere. Due to the high impact of clouds on shortwave and long wave radiation, mis-identification of cloud pixels as surface or vice versa can significantly affect the quality of any satellite remote sensing product. Traditionally, cloud detection over snow or partially snow-covered areas is a difficult undertaking. For example, the performance of the MOD35 algorithm (Ackerman *et al.*, 2010) declines drastically in the winter seasons over mid-latitude areas when validated by collocated Aqua MODIS/CALIOP data. Complicated land surface conditions, especially when the surface consists of mixed snow and vegetation/soil areas, still pose serious challenges to current threshold-based cloud detection schemes.

We propose a novel machine-learning-based cloud detection scheme to improve the cloud detection over mixed-snow-vegetation areas. Unlike other machine learning based cloud detection methods, which use human-classified satellite data as training data, the cloud/clear-sky identifier in our method is trained by simulated data from extensive radiative transfer calculations for a variety of surface and cloud conditions. By establishing a large training dataset which, incorporates as many different snow, vegetation, soil and mixed cases as possible, we were able to train a neural network classifier and apply it to real satellite images. Figure 1 shows comparisons of cloud mask results obtained by our SCM algorithm and the MODIS cloud mask (MOD35) for Aqua MODIS data over East Asia. It can be seen that our SCM algorithm shows great improvement over MOD35 by correctly identifying clear-sky snow and snow-mixed-vegetation/soil pixels in the circled areas. We tested the performance of our algorithm using long-

[1] Department of Physics, Stevens Institute of Technology, Hoboken, USA
[2] Meteorological Research Institute, Tsukuba, Japan
[3] Japan Aerospace Exploration Agency
[4] Okayama University, Okayama, Japan

term collocated CALIOP/Aqua MODIS measurements, which provide the most reliable assessment of cloud mask algorithms to date, in a way similar to that reported by Chen et al. (2014). The CALIOP 1 km cloud detection results are used as the benchmark and the whole year data of 2008 over areas including East Asia and Europe are used. It can be seen in Figure 2 that our SCM algorithm outperforms MOD35 (employing ten reflectance bands and nine thermal infrared bands) during the winter season with high Hanssen Kuiper Skill scores throughput, using only six satellite channels in the visible and near infrared wavelength range.

More validations of our SCM algorithm are in progress using collocated Aqua MODIS and CALIOP data over more locations such as North America, Greenland and Antarctica.

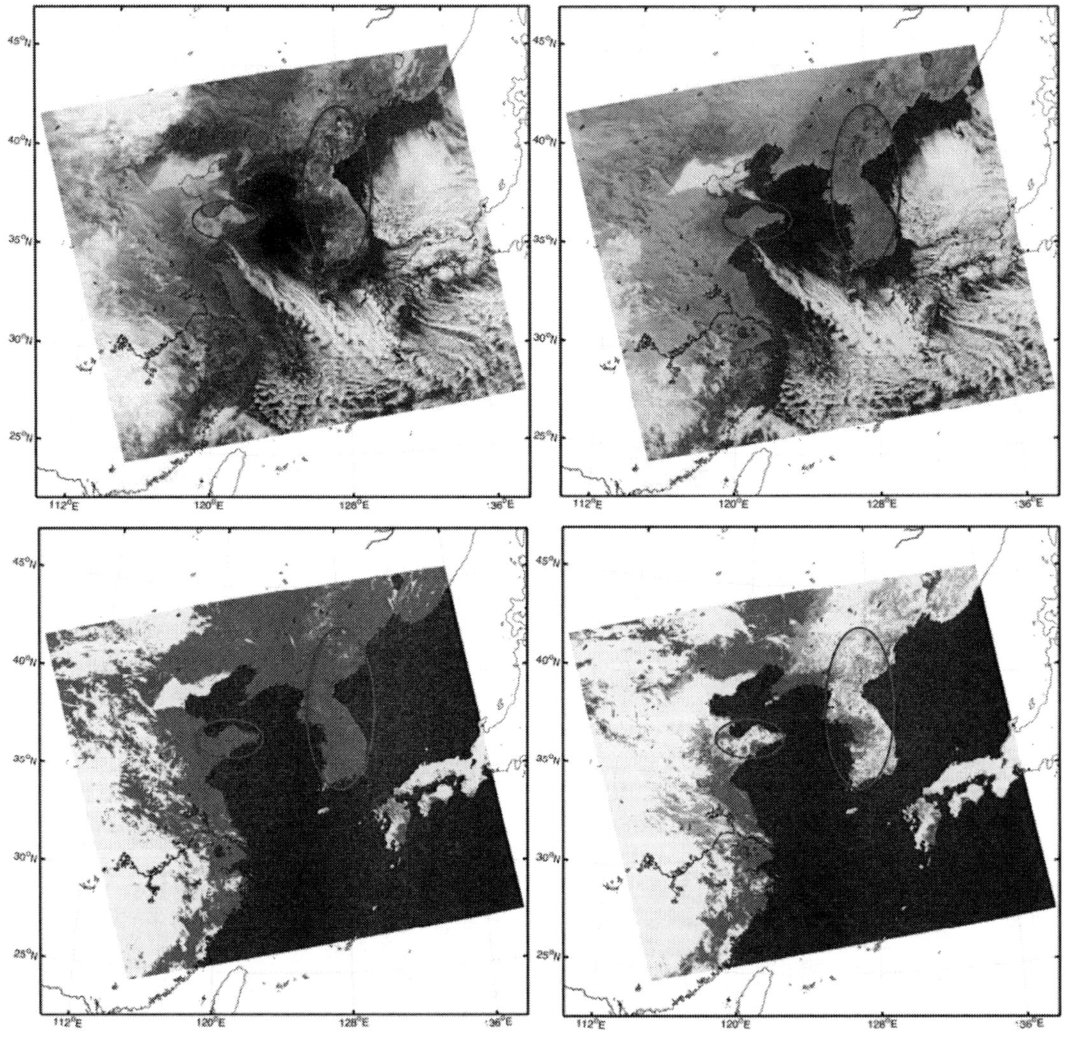

Figure 1. Cloud mask result for a MODIS Aqua image over East Asia, on January 24, 2003. Upper panels: Left: True color RGB; Right: False color RGB (0.65 um, 2.13 um, and 10.8 um brightness temperature); Lower panels: Left: SCM cloud mask; Right: MYD35 C6 cloud mask.

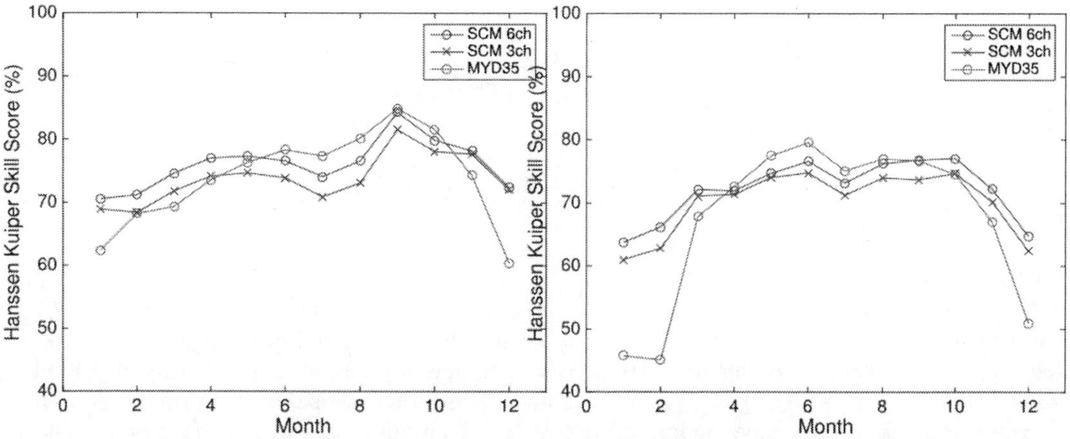

Figure 2. Hanssen Kuiper Skill Score of our cloud mask algorithm (SCM) and MYD35 in 2008 over East Asia (left) and Europe (right). Note the superior performance of the SCM in the winter season.

REFERENCES

Ackerman S, Frey R, Strabala K, Liu Y, Gumley L, Baum B. 1998. Discriminating clear-sky from cloud with MODIS. *Journal of Geophysical Research*, **103**(D24): 32141-32157.

Chen N, Li W, Tanikawa T, Hori M, Aoki T, Stamnes K. 2014. Cloud mask over snow/ice covered areas for the GCOM-C1/SGLI cryosphere mission. *Journal of Geophysical Research*, **119**(D21): 12287-12300.

Snowfall Variability and Change in the Southern Appalachian Mountains, 1910-2017

MONTANA A. ECK[1] & L. BAKER PERRY[1]

ABSTRACT

Climate variability, trends, and associated relationships with global and hemispheric teleconnection patterns in mountainous regions remain poorly understood. Previously identified as an anomalous region with respect to climate change, the southern Appalachian Mountains provide a unique opportunity for analyzing long-term trends and variability in the winter season alongside known large-scale patterns. This paper investigates trends in climatological winter snowfall and analyzes relationships with global and hemispheric teleconnection patterns between 1910 and 2017. Results reveal that the southern Appalachian Mountains have experienced high intraregional variability of trends in snowfall. Statistically significant ($\alpha < 0.05$) relationships were found with snowfall relative to phases of the AO, ENSO, NAO, and PDO. The negative phases of the AO and NAO, the warm phase of the PDO, and El Niño patterns favor positive snowfall anomalies throughout the southern Appalachian Mountains. However, the strength of the relationships between each of these global/hemispheric patterns is spatially diverse with variability of snowfall in the western reaches and high elevations being highly dependent on the phase of the AO and NAO. However, eastward facing slopes and foothill locations see ENSO as a dominant force for snowfall. The analysis of these large-scale patterns reveals favorable scenarios for anomalous snowfall to occur in the region

Keywords: Climate Variability, North Atlantic Oscillation, Pacific Decadal Oscillation, Southern Appalachian Mountains, Snowfall

INTRODUCTION

The southern Appalachian Mountains (SAMs) contain some of the most diverse biological, topographical, and climatological environments in the United States and offer a unique backdrop to investigate winter climate variability and change (Fig. 1). Exhibiting the greatest topographic relief in the eastern US, the mountainous terrain of the SAMs plays a pivotal role in weather patterns, where orographic effects result in high spatial variability in both precipitation and snow cover across the region (Sugg et al., 2016). Annual snowfall can vary substantially with some valley locations averaging less than 25 cm of snowfall each winter and the adjacent mountaintops routinely surpassing 250 cm (Martin et al., 2015). Despite multiple studies investigating long-term climatic changes in the southeastern United States (SEUS) as a whole, few have focused on changes and trends in the SAMs (Ellenburg et al., 2016).

[1] Appalachian State University, Boone, NC

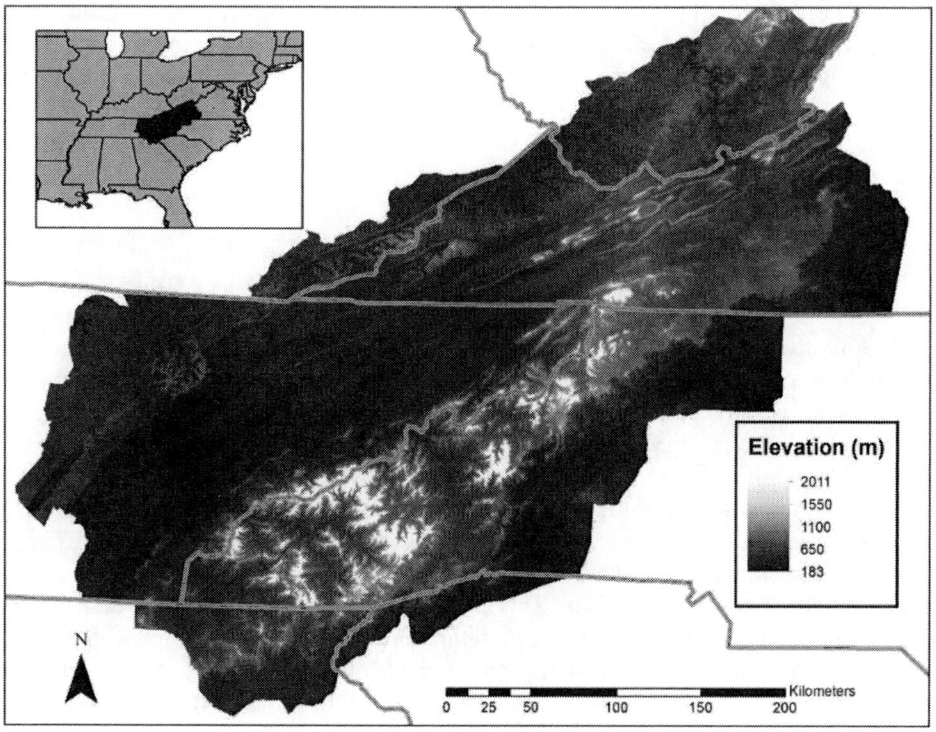

Figure 1: Topography of the southern Appalachian Mountains located primarily in the southeastern United States. Study area spans the states of Georgia, South Carolina, Tennessee, North Carolina, Virginia, Kentucky, and West Virginia.

Identified as globally anomalous climate change regions, the SAMs and SEUS experienced a noticeable cooling trend of mean temperatures in the 20th century (Portmann et al., 2009). Often grouped in studies with surrounding states that experience much less snowfall than the mountains themselves, it can be difficult to understand what sort of trends in snowfall exist in the SAMs. What is known, however, is that the number of snowfall events in the SEUS have been decreasing considerably since 1930 (Kluver & Leathers, 2015). This decline is not limited to only the SEUS, with a majority of North America having experienced decreases in snowfall and snow cover since the start of the 20th century (Kunkel et al., 2016). The number of days with snow cover in the winter season have also experienced a widespread decline in the US, although there are anomalous regions where the length of the snow cover season has increased (Zion et al., 2011).

Due to their location in the mid-latitudes and proximity to the Great Lakes, Gulf of Mexico, and Atlantic Ocean, the SAMs are in an area ideal for the development of snowstorms throughout the winter season. Cyclones that originate in the Gulf of Mexico before tracking along the Atlantic Coast, identified as Miller A and Miller B systems, are the primary contributors to heavy snowfall in the region (Miller, 1946). A variety of other synoptic patterns known to produce snowfall in the SAMs include upper level cut-off lows, clippers, and north-eastward tracking Colorado lows (Perry et al., 2010; Kelly et al., 2012). Spatially, Gulf lows contribute a majority of snowfall to eastern facing slopes. Northwest flow snow (NWFS) events contribute more than 50% of the annual snowfall to windward and high elevation locations (Perry et al., 2010). With the variety of storm patterns affecting the region and the spatial distribution of snowfall on both the event and annual basis being drastically different, analysis of trends in snowfall is crucial to understanding the role global teleconnections play in seasonal variability within the SAMs climatic subregions.

The identification of the SAMs as a globally anomalous region regarding climate change and the known impacts of several global and hemispheric teleconnection patterns guided this research by proposing the following questions: (1) How has the winter climate of the SAMs changed since the early 20th century and (2) what are the influences of large-scale global and hemispheric patterns on the interannual variability of snowfall? In this study, we analyze long-term trends in snowfall across different climatic subregions of the SAMs in order to establish global and hemispheric teleconnection pattern regimes favorable for producing climatically anomalous conditions in the SAMs.

METHODS OF ANALYSIS

Study Area and Climate Regions

Stretching from Georgia to West Virginia and encompassing the Black, Blue Ridge, and Great Smoky Mountains, this study area is known to be one of the most biologically, climatologically, and topographically diverse regions in the United States. Elevations within the study area reach their peak at 2,037 m asl (above sea level) at the top of Mt. Mitchell, the highest point east of the Mississippi River. Snowfall amounts in the SAMs on both the event and annual basis can vary drastically, with these differences being most pronounced in higher elevation and valley locations. Variability of snowfall amounts in the SAMs is enhanced by NWFS, which accounts for 50% of annual snowfall on windward and higher elevation locations but contributes little to eastward facing slopes and lower elevation locations (Perry et al., 2007). It is due to these climatic differences that the study area has been divided into 14 unique climate regions based on similarities in snowfall patterns, elevation, and topography (Fig. 2) (Perry, 2006; Sugg et al., 2016). A final super region, simply identified as the SAMs, takes the snowfall records from the 12 regions with adequate data to understand trends on a larger regional scale. The technique of averaging each of the climate region's COOP stations seasonally helps to alleviate some of the issues that would be prevalent with missing or inaccurate data at individual COOP stations (Table 1).

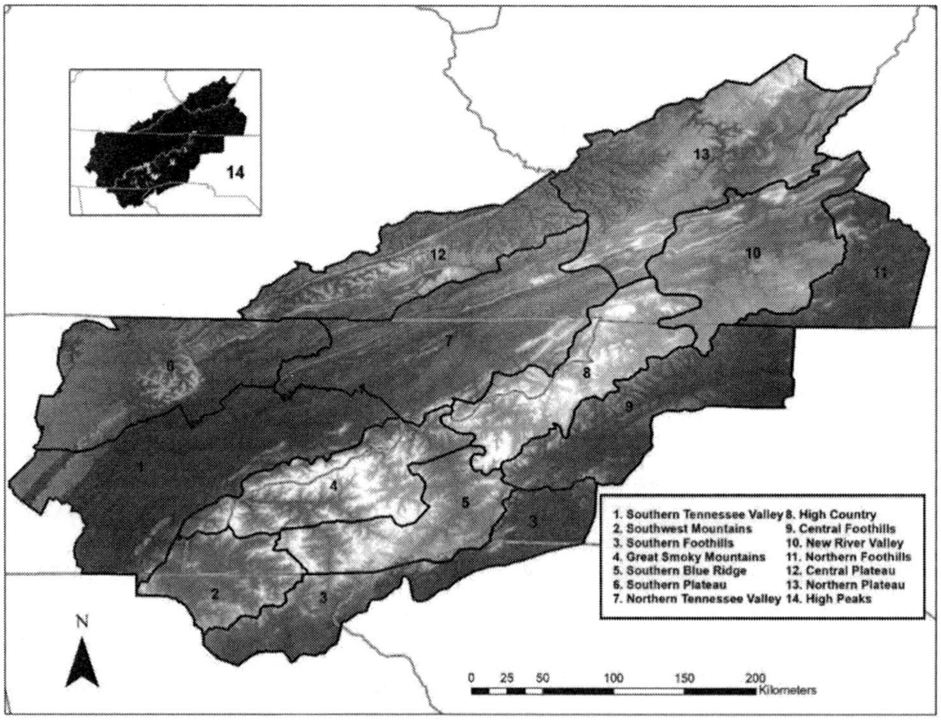

Figure 2. Climatic subregions of the southern Appalachian Mountains.

Table 1. Climatic subregions and associated observer stations.

	Climate Region	COOP Stations
1.	Southern Tennessee Valley	Athens, Cleveland, Dayton, Kingston, Knoxville, Lenoir City, Pikeville
2.	Southwest Mountains	Blairsville, Andrews, Murphy, Copperhill
3.	Southern Foothills	Clayton, Helen, Toccoa, Forest City, Tryon, Pickens
4.	Great Smoky Mountains	Hot Springs, Marshall, Waynesville, Gatlinburg
5.	Southern Blue Ridge	Asheville, Brevard, Cullowhee, Fletcher, Franklin, Highlands
6.	Southern Plateau	Allardt, Crossville, Norris, Oneida, Tazewell
7.	Northern Tennessee Valley	Bristol, Elizabethton, Greeneville, Kingsport, Rogersville, Abingdon
8.	High Country	Banner Elk, Blowing Rock, Boone, Celo, Jefferson, Transou, Mountain City
9.	Central Foothills	Lenoir, Marion, Morganton, North Wilkesboro, Yadkinville
10.	New River Valley	Sparta, Blacksburg, Floyd, Galax, Pulaski, Wytheville
11.	Northern Foothills	Martinsville, Philpott Dam, Roanoke, Rocky Mount, Stuart
12.	Central Plateau	Baxter, Pennington Gap, Wise, Big Stone Gap, Grundy
13.	Northern Plateau	Burkes Garden, Bluefield, Flat Top, Lewisburg, Union, White Sulphur Springs
14.	High Peaks (Above 1220 m)	Grandfather Mountain, Mt. Mitchell, Swannanoa, Mt. Leconte

Climate Data

This study defines the winter season as the climatological winter months of December, January, and February. Past research has shown that the signal between snowfall and global/hemispheric patterns is strongest during this period and weakens once the transition to spring occurs (Hartley, 1999). The analysis period for this study spans from December 1910 to February 2017. For each of the climate regions, monthly snowfall data for individual COOP stations were collected from the National Center for Environmental Information's Global Historical Climatology Network. Data from these stations were collected by volunteer observers, which poses concern regarding the validity of some datasets, with the considerable challenges in accurately measuring snowfall and observer error often calling into question snowfall totals (Doesken & Leffler, 2000). In order to maintain as accurate of a representation as possible, this study excluded data from stations with less than 90% coverage of the boreal winter season in the observations. The mean snowfall for each region was calculated by averaging the total snowfall from each of the region's COOP stations that met the 90% threshold.

Identified as recurring and persistent large-scale patterns of pressure and circulation, global and hemispheric teleconnection patterns have a profound impact on weather conditions regionally. For this study, we identified two dominant Pacific patterns and three Arctic/Atlantic patterns. Gathered from NOAA's Earth System Research Laboratory (ESRL), the winter indices of the El Niño-Southern Oscillation (ENSO) and Pacific Decadal Oscillation (PDO) patterns were obtained by averaging the DJF monthly indices for each teleconnection. A seasonal value of ENSO found to be $\leq$ -0.5 implied that the winter season was associated with a La Niña pattern, values $\geq$ 0.5 indicated a boreal winter pattern associated with El Niño, and values that fell between -0.5 and 0.5 indicated a neutral phase of ENSO. Using SST anomaly data poleward of 20N in the Pacific basin, the average DJF index of the PDO represents two distinct phases. Positive DJF indices for the PDO indicated a warm phase for that season and a negative value indicated a cool phase. The three identified Arctic/Atlantic teleconnection patterns analyzed in this study include the Atlantic Multidecadal Oscillation (AMO), Arctic Oscillation (AO), and NAO. For the purpose of this study, any averaged negative values found in a given season for the AO and NAO translates to a seasonal pattern dominated by the negative phase of the teleconnection, and a positive value implies the inverse (NCAR, 2016; NCAR, 2017). Also obtained from NOAA's Earth System Research Laboratory, positive values of the AMO indicate a warm phase of the pattern, with negative values indicating the cool phase for that winter season.

All of the data were tested for normality using the Anderson-Darling test (α<0.05). In order to analyze differences in snowfall based on each teleconnection pattern, we used the Mann-Whitney U two-sample rank sums test, which is a nonparametric test that examines differences of means. Using simple Pearson's correlation, we analyzed the relationship of snowfall with respect to the identified global and hemispheric patterns for each climate subregion. P-values for this stage of the project were set at the 0.05 level to assure the most accurate results possible. After establishing the strongest relationships between the identified global teleconnection patterns and climate conditions in the SAMs, the study also grouped similar teleconnection patterns in order to identify conditions favorable for producing anomalous climatic patterns in the study area.

RESULTS AND DISCUSSION

Trends in Snowfall

The SAMs have seen no significant trend in snowfall since the start of the 20th century. While some subregions do show a general decline in snowfall over time, a majority of the SAMs have experienced a non-significant increase, namely the High Peaks. Those found along the eastern escarpment have seen a non-significant decline in snowfall (Fig. 4). It is important to note that snowfall amounts are also spatially variable with regions that border each other often exhibiting drastically different snowfall totals. For example, the 2009-2010 winter season saw the High Country subregion accumulate 146.5 cm of snow, while the Northern Tennessee Valley and the Central Foothills only saw 34.1 cm and 53.7 cm respectively. For the SAMs as a whole, the 1917-1918 winter season experienced the greatest amount of snowfall with a mean of 70.0 cm, which is substantially more than the 1.1 cm experienced during 1931-1932, the winter season with the least amount of snowfall. Snowfall in the 1960s and early 1970s was highly anomalous, with snowfall between December 1962 and February 1971 averaging more than 40.4 cm across the SAMs. This is in stark contrast to the start of the 2000s, where from December 1999 to February 2008 snowfall in the SAMs averaged 17.5 cm, with eastern subregions like the Southern Blue Ridge and Central Foothills averaging less than half of their 1951-1980 normal snowfall.

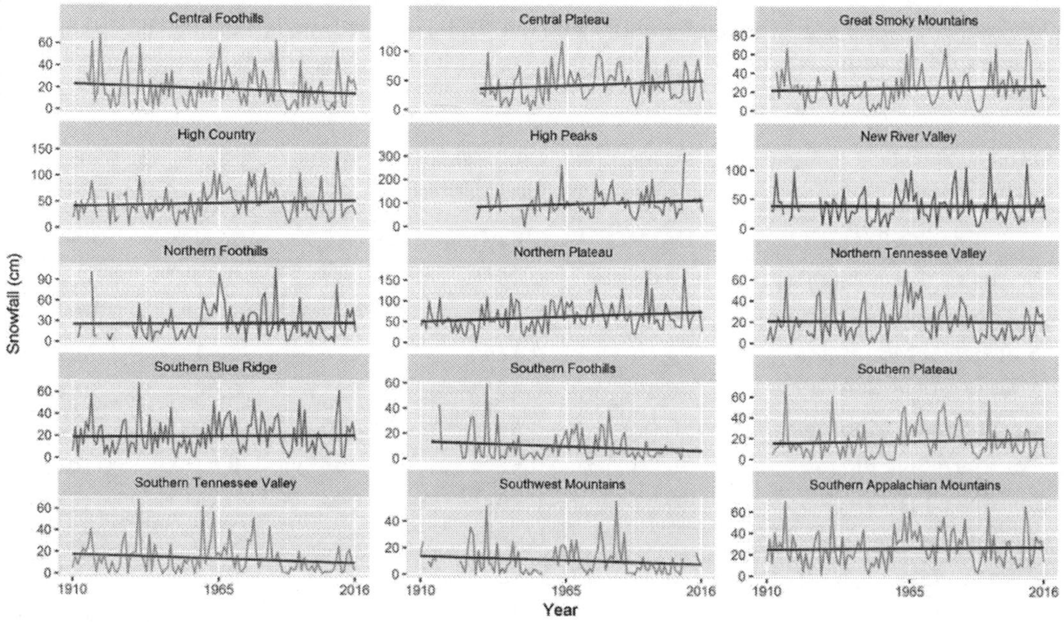

Figure 4. Trends in snowfall for the subregions of the southern Appalachian Mountains.

While no long-term trend can be detected with snowfall, recent decades have hinted at a general decline in snowfall totals. One explanation could be unseen changes in the lower troposphere, where the presence and strength of the warm nose can often mean the difference between snow and rain in the SAMs (Perry et al., 2010). If temperatures in the warm nose rise above 0°C, snowflakes are more likely to melt, reaching the surface as a mixture of sleet and freezing rain (Bell & Bossart, 1988). Highlighting the importance of understanding subregional patterns in the SAMs, there are some cold air damming events where the topography of the southeastern slopes and adjacent foothills assists in locking in cold air, allowing frozen precipitation to persist much longer than in areas to the south and northwest. Further analysis is needed to understand how synoptic patterns and storm tracks have changed during the period and how these may be related to the snowfall trends. Slight deviations in storm track, frequency, and intensity often can mean the difference between rain and snow in the SAMs.

Global and Hemispheric Pattern Impacts on Snowfall

The results of the Mann-Whitney tests indicate that for all 14 subregions within the SAMs, the negative phase of the NAO patterns are significantly ($\alpha<0.05$) snowier than positive NAO winters. However, differences with the AO are not as significant with eastern subregions, like the Northern Foothills and New River Valley. In regards to the PDO, similar results were found with a majority of the subregions experiencing significantly ($\alpha<0.05$) snowier conditions during the warm phase of the pattern, with those found east of the Blue Ridge Escarpment having the least significant differences. Inversely, only eastward regions were found to have significant differences in snowfall dependent on the ENSO phase. No significant differences in snowfall were found based on the phase of the AMO.

For each of the 14 climate regions analyzed in this study, the AO, ENSO, NAO, and PDO display significant relationships with boreal winter snowfall (Table 4). However, the strength of these relationships are very much spatially dependent. Subregions prone to receiving NWFS have much stronger relationships with the AO and NAO than eastward facing slopes. Similarly, the PDO has a much stronger relationship with western subregions. The PDO, while still significant for a majority of the SAMs, has much weaker relationships regarding snowfall with eastward facing slopes and foothill subregions. ENSO is found to favor three Gulf low and Miller system favored subregions, the Southern Blue Ridge, Central Foothills, and the High Peaks. Westward facing subregions have no relationship with ENSO regarding snowfall. The AMO lacks any significant relationships with snowfall in the SAMs.

Table 4. Statistical relationships between climatic subregion snowfall and identified large-scale global and hemispheric patterns.

Climate Regions	DJF AMO	DJF AO	DJF ENSO	DJF NAO	DJF PDO
S. Appalachian Mountains	-0.09	**-0.46****	0.24*	**-0.55****	**0.29****
S. Tennessee Valley	-0.06	**-0.46****	0.13	**-0.52****	**0.30****
Southwest Mountains	-0.16	**-0.42****	0.19	**-0.41****	**0.28****
Southern Foothills	-0.10	**-0.39****	0.23*	**-0.40****	**0.22****
Great Smoky Mountains	-0.07	**-0.34****	0.20*	**-0.49****	0.14
Southern Blue Ridge	-0.04	**-0.42****	**0.25****	**-0.51****	**0.26****
Southern Plateau	-0.11	**-0.39****	0.09	**-0.46****	**0.27****
N. Tennessee Valley	-0.07	**-0.48****	0.11	**-0.54****	**0.26****
High Country	-0.09	**-0.50****	0.23*	**-0.58****	**0.26****
Central Foothills	-0.14	**-0.37****	**0.28****	**-0.41****	0.18
New River Valley	-0.09	**-0.27****	0.24*	**-0.37****	0.24*
Northern Foothills	-0.09	**-0.27****	0.21*	**-0.35****	0.21*
Central Plateau	-0.19	**-0.35****	0.13	**-0.47****	**0.37****
Northern Plateau	-0.06	**-0.36****	0.20*	**-0.45****	**0.31****
High Peaks	-0.04	**-0.45****	**0.34****	**-0.47****	**0.37****

*Significant at the .05 Level
****Significant at the .01 Level**

The NAO, which is significant ($\alpha<0.05$) for all regions, has the most prevalent relationships with the westernmost mountains and loses strength east of the Blue Ridge Escarpment (Fig. 6). The general relationship between the NAO and snowfall is also shared with the AO, with prevalence for more significant relationships in NWFS event favored areas. For the SAMs as a whole, the 10 snowiest years on record were accompanied by a –NAO phase. The inverse is true for the 10 least snowy winters on record where 90% of the seasons correspond to a +NAO phase. While the +PDO may favor heavier snowfall in the SAMs, the same cannot be said for highly anomalous snowfall as the top ten snowfall years in the region are equally represented by –PDO and +PDO patterns. Similarly, a scattering of strong/moderate El Niño and La Niña events, as well as neutral patterns can be seen in both the top and bottom ten snowfall years.

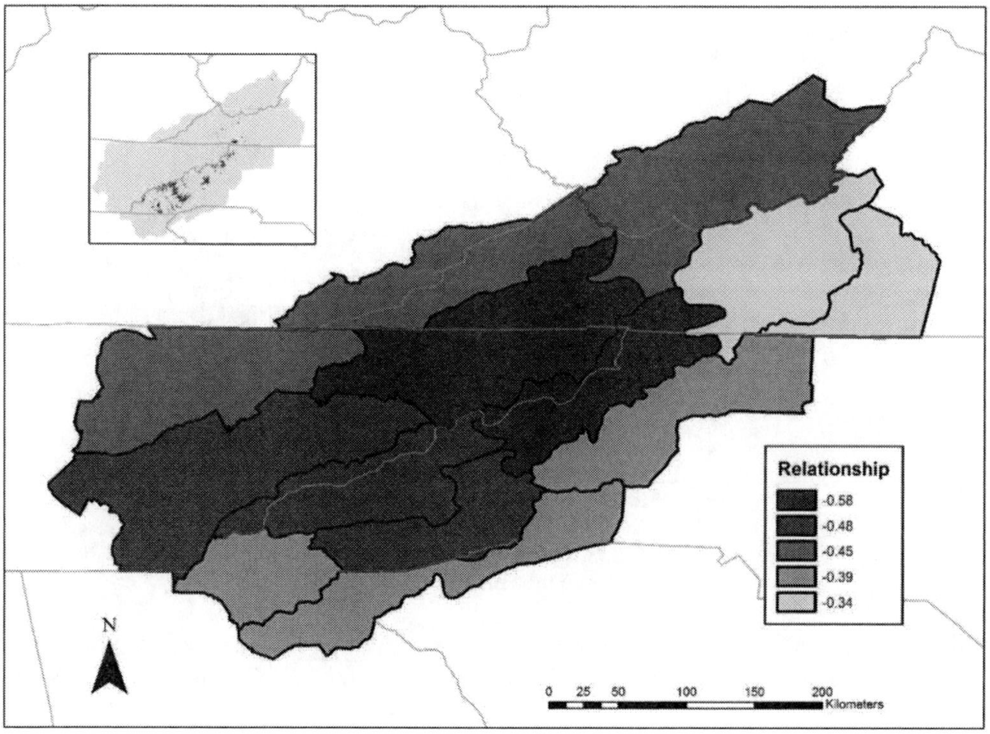

Figure 6. Strength of the relationship between the NAO and snowfall in the southern Appalachian Mountains by climatic subregion.

ENSO and the PDO

Based on the classifications outlined by Jia & Ge (2017), this study identified 13 moderate El Niño events and 12 moderate La Niña events. The study also identified 19 strong El Niño events and 20 strong La Niña events. In the SAMs, there are no substantial differences in snowfall totals during strong El Niño events dependent on the phase of the PDO, with strong El Niño events associated with the –PDO averaging 0.8 cm more than strong El Niño/+PDO events. However, the difference in snowfall amounts with moderate El Niño events are substantial in the SAMs with moderate El Niño events associated with +PDO phases nearly doubling the snowfall amounts found in the region with average seasonal totals of 41.2 cm compared to the 21.0 cm found during moderate El Niño/–PDO patterns (Fig. 8). These findings are evident throughout all 14 climate regions, with the greatest difference existing between snowfall totals found along the western subregions and southernmost extent of the SAMs, with the Southern Tennessee Valley seeing 4.1 times as much snowfall during +PDO/moderate El Niño events than –PDO/moderate El Niño seasons. While still substantially different, subregions along the eastern slopes of the SAMs experience between 1 to 1.5 times as much snowfall during the favorable pattern.

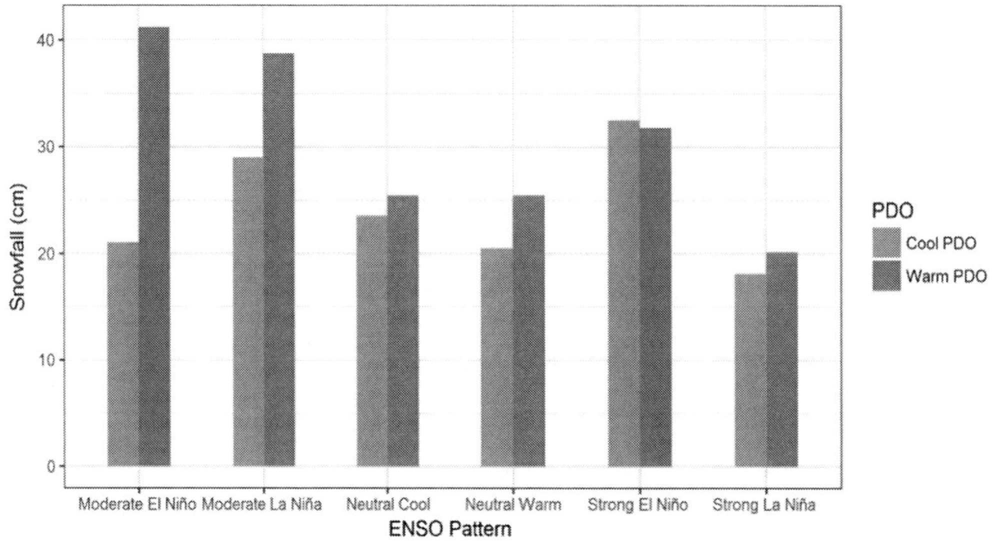

Figure 8. Comparison of snowfall amounts for the southern Appalachian Mountains based on the strength of ENSO and the phase of the PDO.

Little difference in snowfall amounts were found during strong La Niña events. While snowfall totals were below average, strong La Niña events coupled with –PDO patterns average 20.2 cm compared to strong La Niña seasons paired with a +PDO pattern averaging 18.1 cm, both substantially less than that experienced during strong El Niño events. Notably, the comparison of moderate La Niña events dependent on the phase of the PDO is less impressive than that of moderate El Niño events. Moderate La Niña winters coupled with a +PDO pattern amount to 1.3 times the amount of snowfall than moderate La Niña winters coupled with a –PDO. Similar to moderate El Niño events, there is a spatial component to differences in snowfall based on moderate La Niña winters, with consistent 1 to 1.5 times the amount of snowfall during +PDO events throughout western and high elevation subregions, with the differences decreasing substantially in the eastern foothills. Subregions found along the southernmost extent of the study area, such as the Southwest Mountains and Southern Foothills actually receive substantially less snowfall during moderate La Niña/+PDO winters.

Global and hemispheric teleconnection patterns have a considerable impact on the winter climate of the SAMs, with the ENSO, NAO, and PDO having the greatest influences on snowfall throughout the region. Our results confirm the findings and inferences of previous studies that snowfall in the SAMs exhibit statistically significant relationships with the NAO (Fig. 9) (Hartley, 1999). More importantly, our results indicate that the strength of the ENSO pattern and the associated PDO phase are equally important in determining snowfall amounts in the SAMs. Notably, higher snowfall amounts are seen in the region during both moderate phases of the ENSO when coupled with a warm PDO pattern. These differences are evident throughout the SAMs, with the greatest differences seen along the eastern escarpment of the Blue Ridge Mountains.

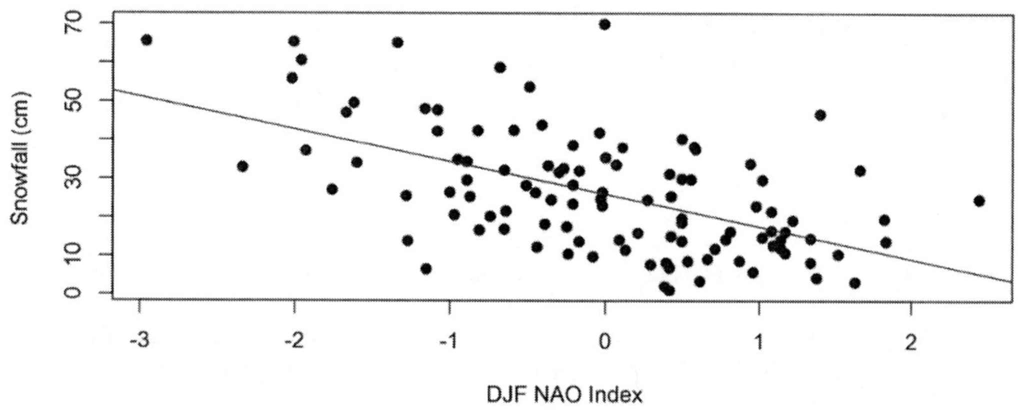

Figure 9. Plotted relationship between snowfall in the SAMs and the winter NAO index.

While past research has identified similar relationships, our study reveals an important spatial aspect to these relationships that previously has not been discussed. Like trends in snowfall, the relationship between each teleconnection pattern and snowfall is spatially diverse. Eastward facing slopes and foothill locations favor Pacific patterns in determining seasonal snowfall amounts, with ENSO playing a much more substantial role in these areas. High elevation and western subregions are more dependent on the pattern of the NAO/AO, with much less Pacific influence. One explanation for these spatial differences are the synoptic patterns commonly associated with snowstorms in the SAMs. For the eastern subregions and foothill locations, Miller systems are the dominant contributor to snowfall amounts and are much more prominent during El Niño patterns. While western subregions and high elevation locations also benefit from Miller storms, a majority of these areas' annual snowfall come from NWFS events (Perry et al., 2007). The negative phase of the NAO is believed to enhance winter troughs along the East coast of the US and in turn creates an environment suitable to produce NWFS off the Great Lakes (Cohen et al., 2014). Due to the eastern foothill locations dependence on Pacific patterns for producing snowfall and high elevation and western subregions favoring Arctic patterns, simply creating seasonal forecasts on the projected pattern of ENSO is faulty in nature.

SUMMARY AND CONCLUSIONS

In this study, long-term regional climate data were analyzed alongside several global and hemispheric teleconnection patterns to improve understanding of the winter climate in the anomalous SAMs. Analysis revealed little to no change in seasonal snowfall since 1910. However, recent years and decades have hinted at more volatility in the winter season, with a general decline in annual snowfall since the mid-1960s throughout the study area. There are several factors that have aided in the region's winter variability. Winter seasons dominated by the negative phase of the AO and NAO favor heavier snowfall for the region with the inverse being true for years with stronger positive phases of the NAO. The Pacific also plays a role with snowfall totals in the region with the warm phase of the PDO corresponding to more snowfall in the SAMs. For ENSO, El Niño events favor positive snowfall anomalies but the relationship is not as spatially extensive as with the AO, NAO, and the PDO. The multidecadal nature of the AMO hindered seasonal analysis, with no relationships being found between the pattern and the winter climate of the SAMs. Continued analysis of climatically anomalous regions like the SAMs is crucial to understanding the impact of naturally occurring global and hemispheric forcings in masking, or in some cases, amplifying the impacts of the warming global climate.

ACKNOWLEDGEMENTS

The authors are grateful to Dr. Douglas Miller, Dr. Peter Soulé, Dr. Johnathan Sugg, Lauren Andersen, Burke McDade, and Zachary Osborne for providing guidance and support. The authors would also like to recognize the Southeast Regional Climate Center and its sister climate centers for distributing useful climatic information to both the researchers and the public. Lastly, the authors would like to recognize the Appalachian State University Office of Student Research for providing conference travel funding.

REFERENCES

Bell GD, Bosart LF. 1988. Appalachian cold-air damming. *Monthly Weather Review*, Vol. **116**, 137–161.

Cohen J, Foster J, Barlow M, Saito K, Jones J. 2010. Winter 2009-2010: A case study of an extreme Arctic Oscillation event. *Geophysical Research Letters* **37**(17). DOI: 10.1029/2010gl044256.

Cohen J, Screen JA, Furtado JC, Barlow M, Whittleston D, Coumou D, Francis J, Dethloff K, Entekhabi D, Overland J, Jones J. 2014. Recent Arctic amplification and extreme mid-latitude weather. *Nature Geoscience* **7**(9): 627–637. DOI: 10.1038/ngeo2234.

Doesken NJ, Leffler RJ. 2000. Snow FOOLIN': Accurately Measuring Snow Is An Inexact But Important Science. *Weatherwise* **53**(1): 31–37.

Ellenburg WL, Mcnider RT, Cruise JF, Christy JR. 2016. Towards an Understanding of the Twentieth-Century Cooling Trend in the Southeastern United States: *Biogeophysical Impacts of Land-Use Change. Earth Interactions* **20** (18):1–31.

Hartley S. 1999. Winter Atlantic Climate and Snowfall in the South and Central Appalachians. *Physical Geography*, **20**(1), 1–13.

Hurrell JW, Deser C. 2009: North Atlantic climate variability: The role of the North Atlantic Oscillation. *Journal of Marine Systems*, **78**, No. 1, 28-41 6–603. DOI: 10.1038/nclimate2583.

Jia X, Ge J. 2017. Modulation of the PDO to the relationship between moderate ENSO events and the winter climate over North America. *International Journal of Climatology*. DOI: 10.1002/joc.5083.

Kelly G, Perry LB, Taubman B, Soulé P. 2012. Synoptic classification of 2009–2010 precipitation events in the southern Appalachian Mountains, USA. *Climate Research* **55** (1):1–15.

Kluver D, Leathers D. 2015. Regionalization of snowfall frequency and trends over the contiguous United States. *International Journal of Climatology* **35**(14): 4348–4358. DOI: 10.1002/joc.4292.

Kunkel KE, Robinson DA, Champion S, Yin X, Estilow T, Frankson RM. 2016. Trends and Extremes in Northern Hemisphere Snow Characteristics. *Current Climate Change Reports* **2**(2): 65–73. DOI: 10.1007/s40641-016-0036-8.

Martin D, Perry LB, Miller D, Soulé P. 2015. Snowfall event characteristics from a high-elevation site in the Southern Appalachian Mountains, USA. *Climate Research* **63**(3): 171–190. DOI: 10.3354/cr01291.

Miller JE. 1946. Cyclogenesis in the Atlantic Coastal Region of the United States. *Journal of Meteorology* **3**: 31–44. .

National Center for Atmospheric Research Staff (Eds). Last modified 18 Aug 2016. "The Climate Data Guide: Hurrell wintertime SLP-based Northern Annular Mode (NAM) Index." Retrieved from https://climatedataguide.ucar.edu/climate-data/hurrell-wintertime-slp-based-northern-annular-mode-nam-index.

National Center for Atmospheric Research Staff (Eds). Last modified 08 Aug 2017. "The Climate Data Guide: Hurrell North Atlantic Oscillation (NAO) Index (PC-based)." Retrieved from https://climatedataguide.ucar.edu/climate-data/hurrell-north-atlantic-oscillation-nao-index-pc-based.

Perry LB, Konrad CE, Hotz DG, Lee LG. 2010. Synoptic classification of snowfall events in the Great Smoky Mountains, USA. *Physical Geography* **31**: 156–171

Perry LB, Konrad CE, Schmidlin TW. 2007. Antecedent Upstream Air Trajectories Associated with Northwest Flow Snowfall in the Southern Appalachians. *Weather and Forecasting* **22**: 334–352.

Perry LB, Konrad CE, Hotz DG, Lee LG. 2010. Synoptic classification of snowfall events in the Great Smoky Mountains, USA. *Physical Geography* **31**: 156–171

Portmann RW, Solomon S, Hegerl GC. 2009. Spatial and seasonal patterns in climate change, temperatures, and precipitation across the United States. *Proceedings of the National Academy of Sciences* **106** (18):7324–7329.
Sugg JW, Fuhrmann CM, Perry LB, Hall DK, Konrad CE. 2016. Sub-regional snow cover distribution across the southern Appalachian Mountains. *Physical Geography* **38**(2): 105–123. DOI: 10.1080/02723646.2016.1162020.
Zion MS, Pradhanang SM, Pierson DC, Anandhi A, Lounsbury DG, Matonse AH, Schneiderman EM. 2011. Investigation and Modeling of winter streamflow timing and magnitude under changing climate conditions for the Catskill Mountain region, New York, USA. *Hydrological Processes* **25**(21): 3289–3301. DOI: 10.1002/hyp.8174.

Contribution of Lake-Effect Snow to the Catskill Mountains Snowpack

DOROTHY K. HALL,[1,2] NICOLO E. DIGIROLAMO,[3] ALLAN FREI[4]

ABSTRACT

Meltwater from snow that falls in the Catskill Mountains in southern New York contributes to reservoirs that supply drinking water to approximately nine million people in New York City. Using the NOAA National Ice Center's Interactive Multisensor Snow and Ice Mapping System (IMS) 4km snow maps, we have identified at least 32 lake-effect (LE) storms emanating from Lake Erie and/or Lake Ontario that deposited snow in the Catskill/Delaware Watershed in the Catskill Mountains of southern New York State between 2004 and 2017. This represents a large underestimate of the contribution of LE snow to the Catskills snowpack because many of the LE snowstorms are not visible in the IMS snow maps when they travel over snow-covered terrain. Most of the LE snowstorms that we identified originate from Lake Ontario but quite a few originate from both Erie and Ontario, and a few from Lake Erie alone. Using satellite, meteorological and reanalysis data we identify conditions that contributed to LE snowfall in the Catskills. Clear skies following some of the storms permitted measurement of the extent of snow cover in the watershed using multiple satellite sensors. IMS maps tend to overestimate the extent of snow compared to MODerate resolution Imaging Spectroradiometer (MODIS) and Landsat-derived snow-cover extent maps. Using this combination of satellite and meteorological data, we can begin to quantify the important contribution of LE snow to the Catskills Mountain snowpack. Changes that are predicted in LE snowfall from the Great Lakes could impact the distribution of rain vs snow in the Catskills which may affect future reservoir operations in the NYC Water Supply System.

Keywords: lake-effect snow, Catskills, Lake Erie, Lake Ontario, IMS, MODIS

INTRODUCTION

Snowmelt is an important source of water for approximately nine million people in New York City (NYC) as well as others in New York State who rely on the NYC Water Supply System (NYCWSS) for their water needs. The NYCWSS is the largest unfiltered water supply system in the United States (Matonse et al., 2011). On average, runoff emanating from the six basins of the Catskill/Delaware watershed in the Catskill Mountains has supplied 90 percent of NYC's water demands. The westernmost basin, the Cannonsville, contains the second largest reservoir used for NYC drinking water. The contribution of snowfall to total annual precipitation in the Catskill/Delaware Basin has been 20%–30% between ~1950 through ~2010 (Frei et al., 2002; Pradhanang et al., 2011; Anandhi et al., 2011). Some portion of the snowfall in the Catskills emanates from lake-effect (LE) snow from Lake Erie and Lake Ontario.

[1]Terrestrial Information Systems Laboratory, NASA/GSFC, Greenbelt, MD
[2]Earth System Science Interdisciplinary Center/University of Maryland, College Park, MD, dorothy.k.hall@nasa.gov
[3]SSAI, Lanham, MD, nicolo.e.digirolamo@nasa.gov
[4]Department of Geography, Hunter College, City University of New York, NYC, afrei@hunter.cuny.edu

In this paper we examine the contribution of LE snowfall to the Catskills Mountain snowpack in southern New York State using satellite snow-cover maps derived from NOAA's National Ice

Center 4km Interactive Multisensor Snow and Ice Mapping System (IMS), NASA's 500-m resolution MODerate resolution Imaging Spectroradiometer (MODIS) images and standard snow-cover map products, Landsat-derived snow maps, meteorological and reanalysis data.

BACKGROUND

Lake Effect Snowfall in the Eastern Great Lakes Region

During the 20th century, annual snowfall increased across the lake-effect zone of the Great Lakes Basin, likely due to warming of lakes and diminished ice cover (Burnett et al., 2003; Kunkel et al., 2009; Notaro et al., 2013). Specifically, significant increases in LE snow were found between 1951 and 1990 (Norton and Bolsenga, 1993; Leathers and Ellis, 1996; Burnett et al., 2003). However LE snowfall has generally been declining in recent decades in some areas in the Great Lakes Basin (Hartnett et al., 2014) and is projected to decline during the 21st century in part because the air temperature is expected to continue to rise, causing less precipitation to fall as snow (Suriano and Leathers, 2016).

LE snowfall is generated when a cold air mass moves over a warmer lake causing relatively low (3000-m cloud tops) stratocumulus clouds to develop from convective cells (Peace and Sykes, 1966; Pease et al., 1988; Kunkel et al., 2000). Heat and moisture is transferred from the lake to the air. This process is especially effective when the lake is not ice-covered, or at least not fully ice-covered (Norton and Bolsenga, 1993) and when there is a large contrast in temperature (at least 13°C) between the lake surface and the 850 mb air temperatures (Holyroyd, 1971). The air and lake temperatures must have a large temperature contrast for convective transfer of heat from the relatively warm lake surface to the much-colder overlying air. The development and intensity of LE snow depend strongly on the temperature difference between air and water, the extent of ice cover, fetch, and wind direction and speed.

Narrow (5 – 20 km) and elongated (50 – 300 km) snow bands form on the leeward side of lakes (Figure 1), the movement of which is controlled by winds aloft, rather than by surface conditions (Eichenlaub, 1979; Peace and Sykes, 1966; Niziol, 1987; Hartnett, 2013 and Hartnett et al., 2014). In addition to LE snow that forms over a lake, lake-induced snowfall, a term that includes both LE and lake-enhanced snowfall, contributes large amounts of snowfall on the leeward sides of the Great Lakes (Bard and Kristovich, 2012; Suriano and Leathers, 2016). The combination of long, overwater fetch and strong winds can cause narrow bands of precipitating clouds to propagate considerable distances inland (for example, see Niziol et al., 1995). New research by Villani et al. (2017) describes atmospheric parameters that have the greatest influence on the ability of a LE storm to extend inland from Lake Ontario. They concluded that the primary and secondary parameters most-strongly correlated with inland extent of LE snowfall are: 1) the presence of a multi-lake/upstream moisture source connection, and 2) the difference between the lake's surface water temperature and the air temperature at 850 mb.

Lake Erie is the shallowest and the southernmost of the Laurentian Great Lakes with an average depth of only 19 m, resulting in Erie having the greatest amount of ice cover of all of the Great Lakes during most winters. In contrast, Lake Ontario, with an average depth of 85 m, remains ice-free or has only a small percentage of its surface covered by ice during many winters (Wang et al., 2012). Furthermore the east-west orientation of Lake Ontario can allow a long fetch (up to ~300 km) when prevailing northwesterly winds that bring cold, dry air across the relatively warm lake surfaces, are oriented over the long axis of the lake; this long fetch can increase the intensity of the LE storms by allowing the storm to pick up more moisture from the lake.

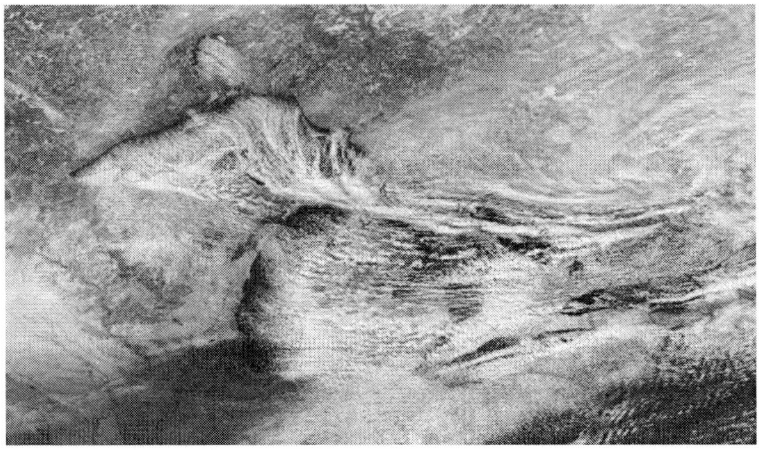

Figure 1. Suomi NPP Visible Infrared Imaging Radiometer Suite (VIIRS) image from 18 November 2014, showing narrow, elongated snow bands over Lake Superior and Lake Michigan / Huron. In this storm the west-southwesterly winds were parallel with the long axis of Lake Erie (which is barely visible below the clouds in the image above) creating a long fetch, allowing the air flowing over the lake to pick up a large amount of moisture. The Cooperative Institute for Meteorological Satellite Studies (CIMSS) Satellite Blog of the University of Wisconsin – Madison [http://cimss.ssec.wisc.edu/goes/blog/archives/17196] reported that cold arctic air with temperatures in the range of about -7 to -4°C flowed across the warm waters (~8-10°C) of Lake Erie and Lake Ontario helping to cause a major LE snowfall event on 18 November 2014. Credit: CIMSS and NASA Earth Observatory.

Up to seven LE synoptic types have been reported to be associated with LE snowfall over Lake Erie and Lake Ontario (Leathers and Ellis, 1996; Suriano and Leathers, 2017). All seven types are associated with low pressure to the north and/or east and high pressure to the west and/or south of Buffalo, NY, and usually with an upper level trough over the United States (Suriano and Leathers, 2017). Irrespective of the origin, the LE snowstorms that garner the most media attention occur when orographic lifting produces a large amount of snowfall that is restricted to a relatively small area on the leeward sides of the lakes. In Figure 2 a time series of IMS 4km snow maps from the 12-13 October 2006 storm that dumped up to 142 cm of snow on parts Buffalo, New York [*https://en.wikipedia.org/wiki/Lake_Storm_%22Aphid%22*] may be seen. The snow did not extend very far away from the shore of Lake Erie, but had a major meteorological and economic impact, though the snow depth was highly variable in the local area.

Heavy LE snow can form in response to surges of cold air from Alberta Clippers that are associated with an anticyclone in the central part of the U.S., especially if in conjunction with a cyclone that is also located over the northeastern U.S. or in the Atlantic Ocean just off of the East Coast of the United States (Notaro et al., 2013). This meteorological situation can produce heavy snow downwind of Lake Erie and Lake Ontario (Niziol, 1987; Ellis and Leathers, 1996; Ballentine et al., 1998). Those storms can also bring moisture and very cold air inland to the Catskills.

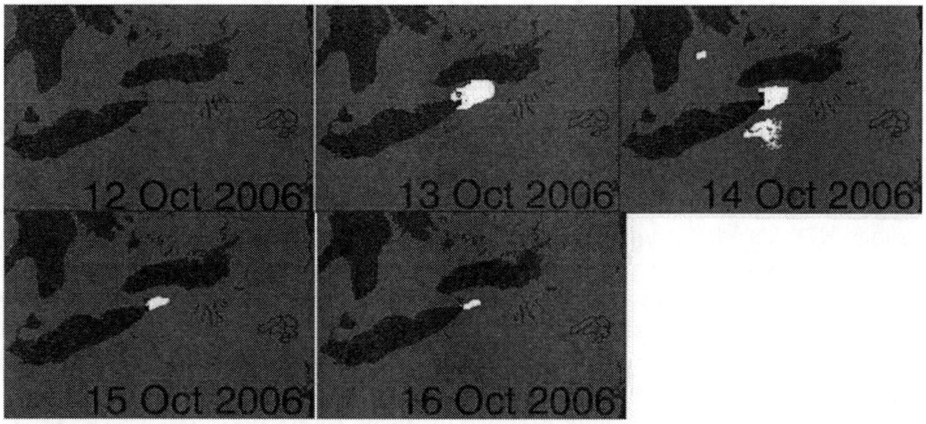

Figure 2. Time series of NOAA IMS 4km resolution snow maps from 12-16 October 2006 showing a LE storm developing and dissipating; a major LE storm dumped up to 142 cm of snow on parts of Buffalo, New York, on 12-13 October 2006. The snow from this storm did not travel very far from the source and did not get close to the Catskill/Delaware Watershed. The six basins of the Catskill/Delaware Watershed are outlined in black.

STUDY AREA

There are six basins of the Catskill/Delaware Watershed: Ashokan, Schoharie, Rondout, Neversink, Cannonsville and Pepacton (Figure 3) that are located up to about 200 km northwest of NYC. Snow conditions can be quite variable within the watershed, both spatially and temporally (Frei et al., 2002; Hall et al., 2016). In particular, the Cannonsville, because of its location, topography and elevation (with elevations ranging from 329 m to 1014 m and a mean elevation 580 m), tends to intercept most of the snow traveling from the west. The Cannonsville Reservoir in the Cannonsville Basin, is about 170 km southeast of Lake Ontario and about 300 km east of Lake Erie.

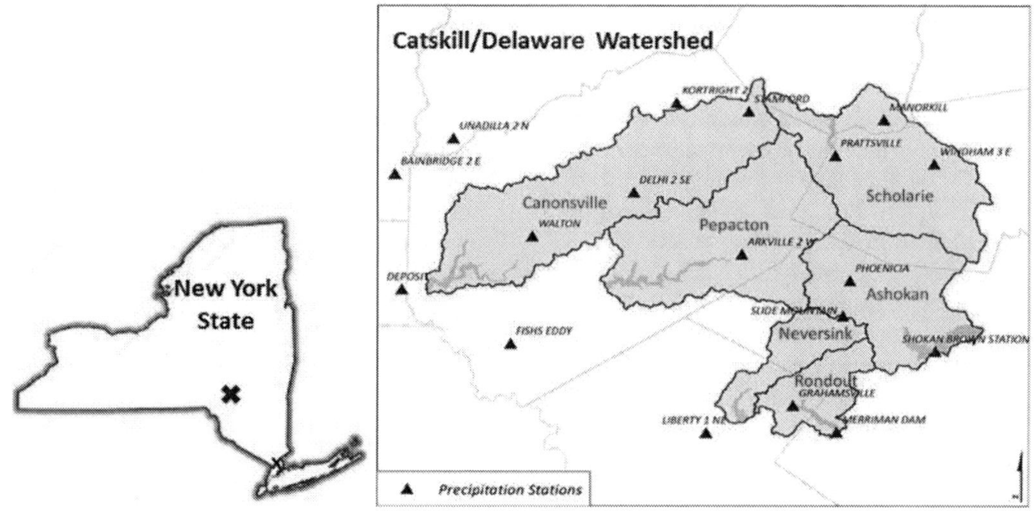

Figure 3. Six basins in the Catskill/Delaware watershed showing locations of meteorological stations (Source: J. Gass, NYC Department of Environmental Protection (NYCDEP) Bureau of Water Supply). The outline of New York State shows the approximate location of the Catskills Mountains (large black X) and of NYC (small black x). Outline of New York State from: Dreamstime.com.

DATA AND METHODOLOGY

Satellite Data

In this paper, we use the following satellite data: 1) NOAA IMS 4km resolution snow maps; 2) MODerate resolution Imaging Spectroradiometer (MODIS) Collection 6 (C6) standard 500-m resolution snow-cover maps; 3) MODIS C6 standard 250-m or 500-m surface-reflectance maps; 4) MODIS C6 standard 1-km land-surface temperature (LST) maps; 5) Landsat-derived snow maps, and 6) Suomi NPP Visible Infrared Imaging Radiometer Suite (VIIRS) imagery.

Multiple satellite and ground station data are utilized to develop the IMS 4km daily SCE maps, so that daily, cloud-cleared snow maps can be provided (Helfrich et al., 2007). Production of the IMS daily snow maps began in 2004. Detection of snow to develop the IMS maps is partially automated and then the maps are finalized manually so that ancillary information, such as might be obtained from meteorological stations, can be in included in the final snow map.

The MODIS standard 500-m resolution daily snow maps, MOD10A1 from the Terra satellite and MYD10A1 from the Aqua satellite, are produced daily, but the ground surface can be fully or partially obscured by cloud cover. In addition, because of the conservative nature of the MODIS cloud mask (Ackerman et al., 1998), there can be clear areas that appear to be cloudy on the daily snow maps. In response to this, a feature of the new C6 MODIS snow maps allows a user to "remove" the cloud mask to visually determine if snow exists beneath the cloud mask. When a clear scene is available, the MODIS snow maps provide an accurate map that may be used to measure the extent of snow cover. Details on the MODIS C6 snow-cover maps may be found in Riggs et al. (2016 and 2017).

MODIS standard surface-reflectance products, MOD09 (Terra) and MYD09 (Aqua) (Vermote et al., 2002) are also used in this work. Additionally the MODIS standard LST product, MOD11_L2 of Wan et al. (2002) was used to estimate lake surface temperatures. Landsat images are used to map SCE in the Catskills using a method similar that used in Hall et al. (2015).

Other Data

Synoptic analyses from NOAA National Centers for Environmental Prediction, Hydrometeorological Prediction Center [*http://www.wpc.ncep.noaa.gov/dailywxmap/index_2005 0105.html*], and Unisys surface data plots [*http://weather.unisys.com/surface/*] were used to study the atmospheric conditions leading to the LE storms identified.

Percent ice concentration and daily lake SWT from Great Lakes Surface Environmental Analysis maps was obtained through the CoastWatch site [*https://coastwatch.glerl.noaa.gov/ statistic/statistic.html*] [*https://coastwatch.glerl.noaa.gov/ftp/glsea/avgtemps/2005/glsea-temps 2005_1024.dat*] operated by NOAA's Great Lakes Environmental Research Laboratory (GLERL). In fact, three methods are evaluated for use in determing the average daily surface water temperature (SWT) of Lake Ontario for the four case studies: MODIS standard LST product, skin temperatures from Modern-Era Retrospective analysis for Research and Applications - 2 (MERRA-2) https://gmao.gsfc.nasa.gov/reanalysis/MERRA-2/_and daily lake average SWT from GLERL's CoastWatch site. MERRA-2 is a reanalysis data product produced at Goddard Space Flight Center with a native resolution of 0.5° latitude x 0.625° longitude x 72 vertical levels (Gelaro et al., 2017). Air temperatures at 850 mb from MERRA-2 were also used. The GLERL SWT is considered to be the most accurate of the three methods.

METHODOLOGY

We inspected each of the IMS daily 4km SCE maps acquired between November and April during each year of the 13-year study period, 2004 – 2017, to identify snow that appeared to emanate from Lake Erie and/or Lake Ontario and was deposited in the Catskills. It was only possible to identify "new" snow, meaning under conditions when there was no snow or very little snow already on the ground immediately prior to the event.

Meteorological conditions were investigated for each suspected LE storm, and in addition, for the four case studies we also used MERRA-2 skin and air temperatures at 850 mb (to compare with LST and daily SWT from GLERL). For the MERRA-2 data the hourly temperatures were averaged for each day, and also skin temperatures from five MERRA-2 pixels covering Lake Ontario were averaged to derive SWT.

The MODIS LST standard product was used to measure SWT in the following manner. A 10 X 10 pixel "box" was selected for Lake Ontario for each day from the clear-sky portions of the lake. Temperatures from both the Terra and Aqua daily products were averaged for the 10 X 10 pixel box to calculate a daily SWT. To provide some quantification of the location and impact of the storm, measurements of SCE from Landsat, MODIS and IMS snow maps were made in the Catskills when possible.

RESULTS

During the 13-year study period 32 suspected LE storms were identified to be emanating from Lake Erie or Lake Ontario (or both lakes) from which snow reached the Catskill Mountains (Table 1). Most of the storms (17) originated from Lake Ontario, six from Lake Erie and the rest came from both lakes. The number of LE snowstorms affecting the Catskills that we identified represents a large underestimate of the number of LE storms that deposited snow in the Catskills during the study period because it is not possible to see the aftermath of snowstorms in the IMS maps developing and moving across the landscape when snow is already on the ground.

We focused on four storms or case studies during the study period: **Case Study 1**: 22 – 24 November 2005, **Case Study 2**: 5 – 10 December 2006, **Case Study 3**: 31 March – 4 April 2013, and **Case Study 4**: 14 – 15 November 2014. Clear skies after the storms enabled MODIS and/or Landsat snow maps to be acquired for three of the four case studies. SCE measured using MODIS and Landsat was in good agreement, while IMS 4km maps tended to overestimate SCE, in part due to the coarser resolution of the IMS as compared to MODIS and Landsat.

We found the expected large (at least 13°C) difference between SWT and air temperature over Lake Ontario for all of the case studies.

Table 1. Dates and day of year (DOY) of suspected lake-effect (LE) storms on which snow reached the Catskill Mountains. The lake from which the storm emanated is also shown. Dates of the four case studies are shown with an asterisk (*).

Date	DOY	Lake
03 Dec 2004	338	Erie
13 Dec 2004	348	Erie then Ontario
05 Jan 2005	005	Both
23 Nov 2005*	327	Ontario
04 Dec 2005	338	Both
19 Jan 2006	019	Ontario
02 Feb 2006	033	Ontario
16 Mar 2006	075	Ontario
05 Dec 2006*	339	Ontario
10 Jan 2007	010	Ontario
17 Jan 2007	017	Ontario
18 Apr 2007	108	Erie
17 Nov 2007	321	Ontario
01 Dec 2007	335	Both?
19-22 Nov 2008	324-327	Both
27 Nov 2010	331	Erie
04 Dec 2010	338	Ontario?
02 Apr 2013*	092	Ontario?

14-15 Nov 2014*	318-319	Ontario?
01 Jan 2015	001	Both?
21 Dec 2015	355	Erie
05 Jan 2016	005	Both
12 Jan 2016	012	Ontario, then Erie
25 Feb 2016	056	Erie
04 Apr 2016	095	Ontario
21 Nov 2016	326	Ontario
29 Nov 2016	334	Ontario
08 Dec 2016	343	Erie
27 Dec 2016	362	Ontario
25 Jan 2017	025	Ontario
28 Feb 2017	059	Ontario
03-05 Mar 2017	062-064	Ontario, then Erie

CASE STUDIES

Case Study 1: 22 – 24 November 2005

Using a time series of IMS 4km snow maps we identified and tracked a massive LE storm emanating from Lake Ontario that deposited snow in the Catskills on 23 – 24 November 2005 (Figures 4 and 5). The banding of snow over Lake Ontario extends inland to the Catskills (Figure 5). According to the IMS snow map, the entire Catskill/Delaware Watershed was snow covered by 23 or 24 November.

Figure 4. Time series (22-24 November 2005) of IMS 4km snow maps showing the progression of a LE storm that deposited snow in the Catskills. The six basins of the Catskill/Delaware Watershed are outlined in black.

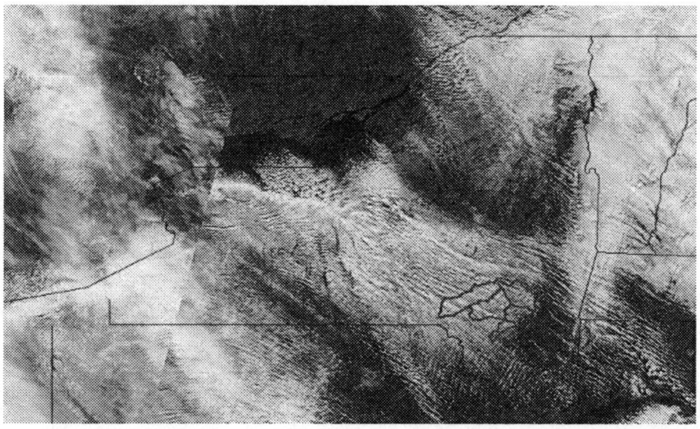

Figure 5. Terra MODIS true-color image of LE snow emanating from Lake Ontario on 23 November 2005. Snow from that storm was deposited in the Catskills. The six basins of the Catskill/Delaware Watershed are outlined in red. Image obtained from NASA's Land, Atmosphere Near real-time Capability for EOS (LANCE).
https://lance.modaps.eosdis.nasa.gov/imagery/subsets/USA4/2005327/USA4.2005327.terra.1km.jpg.

Unisys Surface Data Plots show a cold front from Canada moving in a southeasterly direction on 21-22 November 2005, with winds from the west [*http://weather.unisys.com/surface/*]. LE snow was generated, and deposited overnight and in the morning of November 23rd in the Catskills.

The temperature difference between the SWT of Lake Ontario and the Oswego air temperature increased to 22.3°C by 23 November during the storm (Tables 2a & b). The temperature differences shown in Table 2b are considerably greater than the 13°C, required to spawn LE storms.

Table 2a. Surface water temperature (SWT) of Lake Ontario was determined in three different ways: using land-surface temperature (LST) from MODIS, Tskin from MERRA-2 (M2) and daily SWT from GLERL's CoastWatch site, all shown. Percent ice coverage was zero for all dates.

Date	Avg LST	Daily SWT from CoastWatch	Tskin M2	850 mb Tair
22 Nov 2005	5.0	7.9	6.8	-6.4
23 Nov 2005	5.2	7.7	6.0	-14.6
24 Nov 2005	--	7.6	5.7	-11.5

Table 2b. Differences between GLERL's daily SWT, and MERRA-2 850 mb Tair, in °C.

Date	Temp difference
22 Nov 2005	14.3
23 Nov 2005	22.3
24 Nov 2005	19.1

For comparison, we show the results of three different methods that we used to determine SWT in Table 2a. The most reliable is the measured SWT from GLERL. In Table 2a the temperatures vary by 2.9°C, from a low of 5.0°C using the MODIS LST method, to a high of 7.9°C from GLERL's CoastWatch site.

Case Study 2: 5 – 10 December 2006

Between 5 – 10 December 2006 there was a progression of LE and other storms that contributed to snow deposition in the Catskills. Starting by the morning of 2 December precipitation occurred over Lake Ontario and to the east as determined from Unisys Surface Data Plots. This continued through the evening, but by the next morning (3 December) precipitation had diminished though no snow was evident in the Catskills (Figure 6). On that afternoon, a cold front affected both Lake Erie and Lake Ontario and a low-pressure system traveled over the northern part of the lakes, pulling a front through. Snowfall resumed that evening and continued in the morning of 4 December by which time snow fell from a LE storm emanating from Lake Erie. By the evening of 4 December, winds were from the southwest (over Lake Erie) so LE snow was probably falling by then. Thus for the early part of the event, snow in the Catskills came from Lake Ontario, then a contribution from Lake Erie may have been added.

By 5 December another front deposited still-more snow. During that afternoon, a front was over the Catskills. On the morning of 6 December it was snowing again in the Catskills with some winds from the south so the resulting snowfall was probably not predominately LE snow by that time. Between 5 – 7 December snow is evident in the northwestern part of the Catskill/Delaware Watershed. By 8 December, snow covered most of the watershed's six basins according to the IMS 4km maps, but it is not clear that the more-extensive snow from 8 – 10 December was caused by LE snow alone, though was probably enhanced by moisture from Lake Ontario. Some snow had melted by 10 December as seen on the IMS, Terra MODIS image (Figure 7) and the standard Aqua MODIS C6 snow map (Figure 8). Using a normalized-difference snow index (NDSI) value of 10% or greater, the extent of snow was measured from the MODIS snow map; extent was also measured using the IMS 4km maps. Measurement of SCE was three times greater using IMS as compared to the MODIS snow map (Table 3).

Table 3. Measurement of snow-cover extent in the Catskill/Delaware Watershed using Aqua MODIS and IMS 4km SCE maps, 10 December 2006.

Snow Map	Percent Snow Cover in Watershed	Area of Snow Cover in km^2
Aqua MODIS C6 MOD10A1 snow map	23.7	1004
IMS 4km snow map	71.1	3573

The difference between the SWT of Lake Ontario from GLERL's CoastWatch site and the air temperature at 850 mb from MERRA-2, increased by 4 December during the LE storm (Table 4b) to 20.3°C. The percent ice coverage was zero for each date.

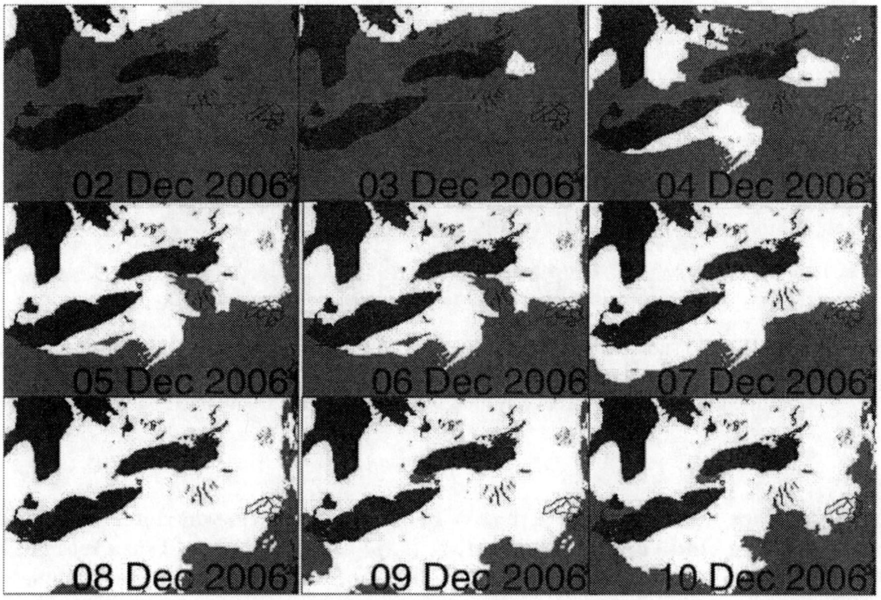

Figure 6. Time series (2 – 10 December 2006) of IMS 4km snow maps showing the progression of the LE storm that deposited snow in the Catskills. The snow that fell in the Catskills originated from a LE storm that appeared to start over Lake Ontario. The six basins of the Catskill/Delaware Watershed are outlined in black.

Figure 7. Terra MODIS 250-m "true-color" image acquired on 10 December 2006. The six basins of the Catskill/Delaware Watershed are outlined in red.

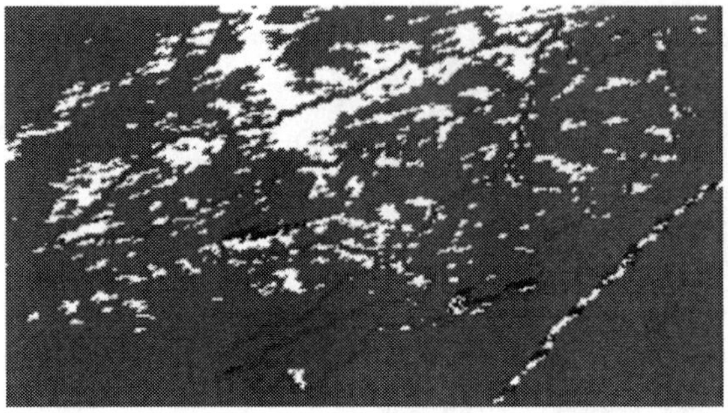

Figure 8. Aqua MODIS C6 snow map from 10 Dec 2006. In this snow map, white represents snow, green represents "no snow," and blue represents water. The six basins of the Catskill/Delaware Watershed are outlined in red. [MYD10A1.A2006344.h12v04.006.2016080151145.hdf]

Table 4a. Surface water temperature (SWT) of Lake Ontario was determined in three different ways: using land-surface temperature (LST) from MODIS, Tskin from MERRA-2 and daily SWT from GLERL's CoastWatch site, all shown. Percent ice coverage was zero for all dates.

Date	Avg LST	Daily SWT from CoastWatch	Tskin M2	Tair 850 mb M2
2 Dec 2006	5.2	7.6	5.9	-10.1
3 Dec 2006	--	7.3	5.3	-8.4
4 Dec 2006	5.7	7.1	4.5	-13.2
5 Dec 2006	5.1	6.9	4.3	-13.2

Table 4b. Differences between daily SWT from GLERL and MERRA-2 850 mb Tair, in °C.

Date	Temp Difference
2 Dec 2006	17.7
3 Dec 2006	15.7
4 Dec 2006	20.3
5 Dec 2006	20.1

On 4 December the SWT estimates from MODIS and MERRA-2 and the measurement from GLERL varied by 2.6°C, from a low of 4.5°C using the MERRA-2 skin temperature, to a high of 7.1°C from GLERL's CoastWatch site (Table 4a).

Case Study 3: 2 April 2013

By the end of March/early April 2013, a series of cold fronts brought cold air to Lake Erie and Lake Ontario from Canada. Winds were west-southwest on 2 April 2013. The IMS 4km SCE maps show the progression of a LE storm that deposited snow in the Catskills in early April 2013 (Figure 9). The Terra MODIS captured a clear view of snow in the Catskill Mountains on 4 April (Figure 10). SCE was measured using different satellite sensors following this storm (Table 5). The Landsat-7 derived snow map and the MOD10A1 snow map provided similar measurements of SCE in the Catskills/Delaware Watershed, while the IMS 4km snow map showed about three times the amount of SCE derived using Landsat-7 data.

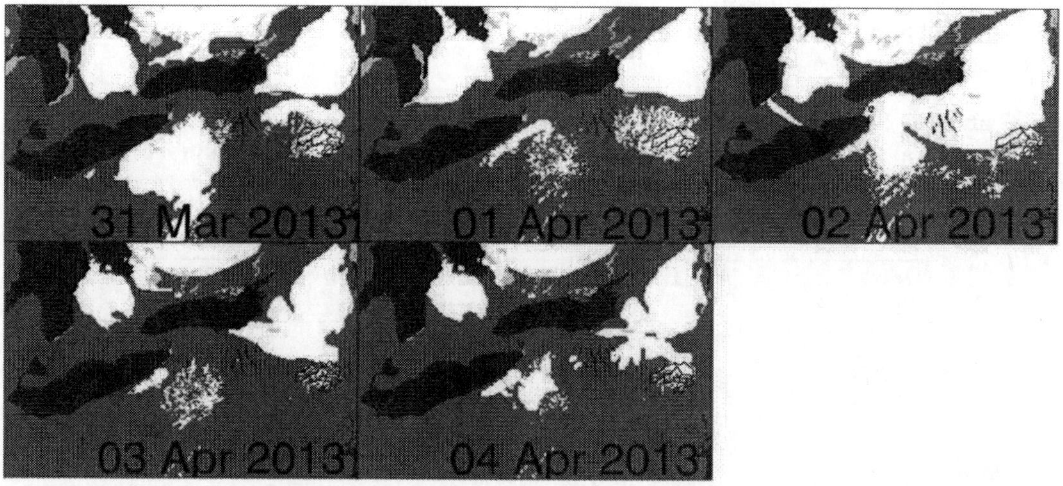

Figure 9. Time series (31 March – 4 April 2013) of IMS 4km snow maps showing the progression of a LE storm that deposited snow in the Catskills. The snow that fell on the Catskills originated from a LE storm that probably started over Lake Ontario. The six basins of the Catskill/Delaware Watershed are outlined in black.

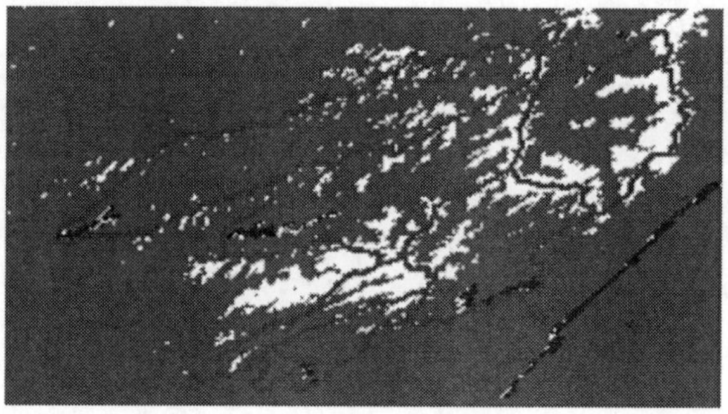

Figure 10. Terra MODIS standard snow map from 4 April 2013. In this snow map, white represents snow, green represents "no snow," and blue represents water. The six basins of the Catskill/Delaware Watershed are outlined in red. [MOD10A1.A2013094.h12v04.006.2016139202830.hdf]

Table 5. Measurement of snow-covered area in the Catskill/Delaware Watershed using Landsat-7 Enhanced Thematic Mapper Plus (ETM+), Terra MODIS and IMS 4km SCE maps, April 2013.

Snow Map/Day Month	Percent Snow Cover in Watershed	Area of Snow Cover in km^2
Landsat-7-derived snow map* / 6 Apr	20.2	829
MODIS C6 MOD10A1 snow map / 4 Apr	24.3	1029
IMS 4km snow map / 4 Apr	49.4	2482

*Derived from: LE70140312013096EDC00

The difference between the SWT of Lake Ontario from GLERL's CoastWatch site and the air temperature at 850 mb from MERRA-2 increased by 2 April during the storm (Table 6b) to 16.8°C. The percent ice coverage was zero for each date.

Table 6a. Surface water temperature (SWT) of Lake Ontario was determined in three different ways: using land-surface temperature (LST) from MODIS, Tskin from MERRA-2 and daily SWT from GLERL's CoastWatch site, all shown. Percent ice coverage was zero for all dates.

Date	Avg LST	SWT from CoastWatch	Tskin M2	Tair M2 850 mb
31 Mar 2013	--	3.1	3.3	0.9
01 Apr 2013	1.6	3.0	2.8	-8.6
02 Apr 2013	1.5	3.0	1.9	-13.8

Table 6b. Differences between daily SWT from GLERL and MERRA-2 850 mb Tair, in °C.

Date	Temp Difference
31 Mar 2013	4.0
01 Apr 2013	11.6
02 Apr 2013	16.8

On 2 April SWT estimates from MODIS and MERRA-2, and the measurement from GLERL varied by 1.5°C, from a low of 1.5°C using the MODIS LST method, to a high of 3.0°C from GLERL's CoastWatch site.

Case Study 4: 14 – 15 November 2014

A lake-effect storm was responsible for copious amounts of snowfall in areas to the east of Lake Erie and Lake Ontario on 14-15 November 2014 (Figure 11). Cool, dry air flowed over the Great Lakes, picking up moisture from the warmer lake surface water, forming cloud streets similar to those seen in the VIIRS satellite image in Figure 1. The snow that fell on the Catskills originated from a LE storm that probably started over Lake Ontario.

A Terra MODIS image was obtained for 19 November 2014 and the study area was enlarged to show detail (Figure 12). Snow was mapped on the same date using the standard MODIS C6 snow map (Figure 13), a Landsat-7-Enhanced Thematic Mapper Plus (ETM+)-derived snow map (Figure 14) and the IMS snow map (not shown). While the Landsat- and MODIS-derived SCE measurements are similar, the IMS 4km maps show almost five times more SCE than was measured using Landsat-7 (Table 7).

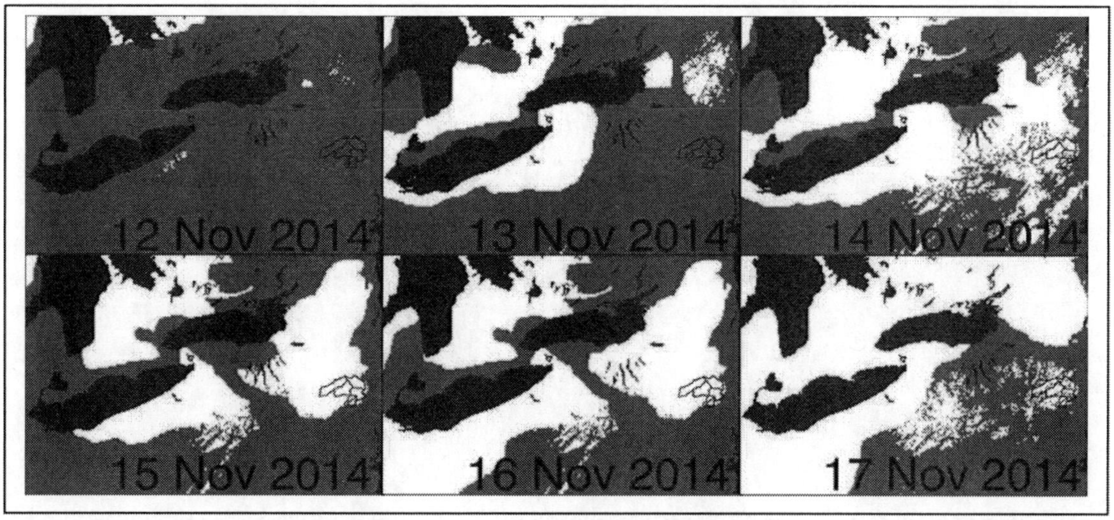

Figure 11. Time series (12 – 17 November 2014) of IMS 4km snow maps showing the progression of the LE storm that deposited snow in the Catskills. The six basins of the Catskill/Delaware Watershed are outlined in black.

Figure 12. MODIS Terra true-color image of the Catskill/Delaware Watershed, 19 November 2014. Source: NASA Worldview.

Figure 13. MOD10A1 C6 NDSI snow map of the Catskill/Delaware Watershed, 19 November 2014. In this snow map, white represents snow, green represents "no snow," and blue represents water. The six basins of the Catskill/Delaware Watershed are outlined in red. [MOD10A1.A2014323.h12v04.006.2016179181758.hdf]

Figure 14. Landsat-7 Enhanced Thematic Mapper Plus (ETM+) – derived snow map Catskill/Delaware Watershed, 19 November 2014. [LE70140312014323EDC00]

Table 7. Measurement of snow-cover extent in the Catskill/Delaware Watershed using different satellite sensors, 19 November 2014.

Snow Map	Percent Snow Cover in Watershed	Area of Snow Cover in km^2
Landsat-7-derived snow map	15.7	647
MODIS MOD10A1 snow map	15.7	664
IMS 4km snow map	65.5	3292

The difference in temperature between the surface of Lake Ontario from GLERL's CoastWatch site and the air temperature at 850 mb from MERRA-2, increased by 15 November during the storm (Table 8b) to 19.7°C. The percent ice coverage was zero for each date.

Table 8a. Surface water temperature (SWT) of Lake Ontario was determined in three different ways: using land-surface temperature (LST) from MODIS, Tskin from MERRA-2 and daily SWT from GLERL's CoastWatch site, all shown. Percent ice coverage was zero for all dates.

Date	Avg LST	SWT from CoastWatch	Tskin M2	Tair 850 mb M2
12 Nov 2014	--	9.4	8.6	-4.4
13 Nov 2014	--	9.0	7.2	-9.7
14 Nov 2014	3.1	8.7	6.8	-10.9
15 Nov 2014	6.8	8.4	6.4	-11.3

Table 8b. Differences between daily SWT from GLERL and MERRA-2 850 mb Tair, in °C.

Date	Temp Difference
12 Nov 2014	13.8
13 Nov 2014	18.7
14 Nov 2014	19.6
15 Nov 2014	19.7

On 15 November the surface water temperature measurements varied by 2.0°C, from a low of 6.4°C using MERRA-2, to a high of 8.4°C from GLERL's CoastWatch site.

Another LE storm occurred on 18 November as reported by the NASA Earth Observatory, and the Cooperative Institute for Meteorological Satellite Studies (CIMSS) Satellite Blog [*http://cimss.ssec.wisc.edu/goes/blog/archives/17196*] but it is unclear from analysis of the IMS snow maps if the snow from that storm that snow reached the Catskills because the ground was already snow covered. See Figure 1 caption for more information.

DISCUSSION AND CONCLUSION

Using a time series of NOAA IMS 4km snow maps, 32 suspected LE storms were identified that deposited snow in the Catskill/Delaware Watershed in the Catskill Mountains during the 13-year study period (2004 – 2017). Most of the storms either originated over Lake Ontario or precipitation was enhanced when air flowed over Lake Ontario, though some of the storms also originated from Lake Erie, or from both lakes. However, we are unable to quantify the contribution of LE storms to the Catskills using satellite methods alone, because it is not possible to track storms that are traveling over already-snow-covered terrain. Also, the visible and near-infrared satellite data provide only snow-cover extent and not snow-water equivalent which would be needed to estimate the amount of snow.

After the snow falls, MODIS and Landsat data allow accurate mapping of SCE when skies are clear. The standard MODIS C6 NDSI snow-cover maps at 500-m resolution provide an excellent way to measure SCE in the Catskills; MODIS measurements of SCE are consistent with measurements made using 30-m resolution Landsat ETM+ derived snow maps. However, the IMS 4km SCE maps tend to greatly overestimate the amount of snow in the Catskill/Delaware Watershed as compared to the Landsat and MODIS maps. The coarser resolution of the IMS is likely responsible for some of the observed overestimation of snow cover.

Lake-effect storms can extend quite far inland and can contribute a large percentage of the total snowfall to inland sites (e.g., Schmidlin, 1992; Villani et al., 2017). Yet there has been some controversy on the extent of the contribution of LE snow to the Catskills because the Catskills are so far inland from the Great Lakes (~170 km from the closest point on the shoreline of Lake Ontario to the Cannonsville Basin, and ~300 km from the closest point on the shoreline of Lake Erie to the Cannonsville Basin). Blechman (1996) reported that LE snow "occasionally" reaches the Catskills depositing only ~0.3 – 1.3 cm of snow per event, yet we show that LE snow makes an important contribution to the Catskills snowpack, especially in some years.

For possible future work on other lakes that might lack daily SWT in-situ measurements as are available from GLERL's CoastWatch site, we investigated the use of two other methods to determine SWT of Lake Ontario for each of the four case studies: determination of SWT from the MODIS LST standard product and from MERRA-2 skin temperature. The measurements varied by up to 2.9°C. The GLERL CoastWatch measurement provided the highest temperature in all cases. We made the assumption that the measured SWT from GLERL's CoastWatch site is the most accurate and thus used the GLERL temperature to compare with the MERRA-2 850 mb air temperature.

All of the measurements of SWT and MERRA-2 air temperature in the case studies showed the expected large (13°C or greater) difference needed to spawn LE snow, with the greatest difference, 22.3°C, found on 23 November 2005 in Case Study 1.

Significance of this work to the New York City Water Supply System and the Catskills

The Cannonsville, which is the westernmost basin of the Catskill/Delaware Watershed, is the second largest reservoir feeding the NYC water supply. It is also the most likely to intercept LE snowfall, and the most vulnerable of the basins to future changes in LE snow patterns because of its location.

Meltwater from Catskills Mountain snowpacks can run off into reservoirs or seep into groundwater aquifers providing important extra storage for the reservoirs that supply water to NYC. Also, a significant portion of the economy in this region is related to winter tourism, so changes in snow cover have economic implications. If the winter circulation changes, and less LE snowfall occurs in the future as predicted (e.g., Notaro et al., 2014; Suriano and Leathers, 2016), this could impact the water availability for NYC and winter tourism in the Catskills.

The projected 21st century temperature increase is likely to affect the Catskills snowpack and the seasonal runoff cycle, resulting in changes in winter vs spring runoff causing reservoir storage levels to increase during the winter (Frei et al., 2002; Matonse et al., 2011). Using a combination of satellite and meteorological data, we can begin to understand the contribution of LE snow to the Catskills Mountain snowpack. In future work, we will focus on methods to quantify the amount of lake-effect snowfall reaching the Catskill Mountains.

ACKNOWLEDGEMENTS

The authors want to thank Dan Leathers and Zachary Suriano of the University of Delaware for their insightful reviews of the paper and for discussions on this work. We also thank George Leshkevich NOAA / Great Lakes Environmental Research Laboratory, Ann Arbor, MI, for discussions about GLERL's CoastWatch data. Richard Cullather / ESSIC – University of Maryland, provided the MERRA-2 skin and air temperature data. Dorothy Hall was funded from a grant, NNX16AD23G, through the Terrestrial Information Systems Laboratory (Miguel Román) at NASA / Goddard Space Flight Center.

REFERENCES

Ackerman SA, Strabala KI, Menzel WP, Frey RA, Moeller CC, Gumley LL. 1998. Discriminating clear sky from clouds with MODIS. *JGR.* **103**(32): 141–32,157, doi:10.1029/1998JD200032.

Anandhi A, Frei A, Pradhanang SM, Zion MS, Pierson DC, Schneiderman EM. 2011. AR4 climate model performance in simulating snow water equivalent over Catskill Mountain watersheds, New York, USA. *Hydrological Processes.* **25**(21): 3302-3311.

Ballentine RJ, Stamm AJ, Chermack EE, Byrd GP, Schleede D. Mesoscale model simulation of the 4–5 January 1995 lake-effect snowstorm. 1998. *Weather and forecasting.* **13**(4): 893-920.

Bard L, Kristovich DAR. 2012. Trend reversal in Lake Michigan contribution to snowfall. *Journal of Applied Meteorology and Climatology.* **51**(11): 2038-2046.

Blechman JB. 1996. A comparison between mean monthly temperature and mean monthly snowfall in New York State. *National Weather Digest.* **20**(4): 41-53.

Burnett AW, Kirby ME, Mullins HT, Patterson WP. 2003. Increasing Great Lake-Effect snowfall during the Twentieth Century: A regional response to global warming? *Journal of Climate.* **16**: 3535-3542.

Eichenlaub VL. 1979. Weather and Climate of the Great Lakes Region, University of Notre Dame Press, 335 pp.

Frei A, Armstrong RL, Clark MP, Serreze MC. 2002. Catskill Mountain water resources: vulnerability, hydroclimatology, and climate-change sensitivity. *Annals of the Association of American Geographers.* **92**(2): 203-224.

Gelaro R, McCarty W, Suarez M and many others. 2017. The Modern-Era Retrospective Analysis for Research and Applications, Version 2 (MERRA-2). *Jour. Climate.* doi:10.1175/JCLI-D-16-0758.1.

Hall DK, Crawford CJ, DiGirolamo NE, Riggs GA, Foster JL. 2015. Detection of earlier snowmelt in the Wind River Range, Wyoming, using Landsat imagery. *Remote Sensing of Environment.* **162**: 45-54, http://dx.doi.org/10.1016/j.rse.2015.01.032.

Hall DK, Frei A, DiGirolamo NE, Porter JH, Riggs GA. 2016. Snow cover and extreme streamflow events in the Catskill/Delaware watershed, New York. *Proceedings of the IAHR*, 31 May – 3 June, 2016, Ann Arbor, MI.

Hartnett JJ, 2013: Spatial and Temporal Trends of Snowfall in Central New York – A Lake Effect Dominated Region, A thesis submitted in partial fulfillment of the requirements for the degree of Master of Science Department of Geography, Environment, and Planning College of Arts and Sciences, 166 pp.

Hartnett JJ, Collins JM, Baxter MA, Chambers DP. 2014. Spatiotemporal snowfall trends in Central New York. Journal of Applied Meteorology and Climatology. **53**(12): 2685–2697, https://doi.org/10.1175/JAMC-D-14-0084.1.

Helfrich S, McNamara D, Ramsay BH, Baldwin T, Kasheta T. 2007. Enhancements to, and forthcoming developments in the Interactive Multisensor Snow and Ice Mapping System (IMS). *Hydrological Processes.* **21**: 1576-1586, doi:10.1002/hyp.6720.

Holyroyd EW III. 1971. Lake-effect cloud bands as seen from weather satellites, *Journal of Atmospheric Science.* **28**: 1165-1170.

Kunkel KE, Wescott NE, Kristovich DAR. 2000. Climate change and lake-effect snow. *Preparing for a Changing Climate: The Potential Consequences of Climate Variability and Change* (2000): 25-28.

Kunkel KE, Palecki M, Ensor L, Hubbard KG, Robinson DR, Redmond K, Easterling D. 2009. Trends in twentieth-century US snowfall using a quality-controlled dataset. *Journal of Atmospheric and Oceanic Technology.* **26**(1): 33-44.

Leathers DJ, Ellis AW. 1996. Synoptic mechanisms associated with snowfall increases on the lee of lakes Erie and Ontario. *International Journal of Climatology.* **16**: 1117-1135.

Matonse AH, Pierson DC, Frei A, Zion MS, Schneiderman EM, Anandhi A, Mukundan R, Pradhanang SM. 2011. Effects of changes in snow pattern and the timing of runoff on NYC water supply system. *Hydrological Processes.* **25**(21): 3278-3288, doi:10.1002/hyp.8121.

Niziol TA. 1987. Operational forecasting of lake effect snowfall in western and central New York. *Weather and Forecasting.* **2**(4): 310-321.

Niziol TA, Snyder WR, Waldstreicher JS. 1995. Winter weather forecasting throughout the eastern United States. Part IV: Lake effect snow. *Weather and Forecasting.* **10**(1): 61-77.

Norton DC, Bolsenga SJ. 1993. Spatiotemporal trends in lake effect and continental snowfall in the Laurentian Great Lakes, 1951–1980. *Journal of Climate.* **6**(10): 1943-1956.

Notaro M, Zarrin A, Vavrus S, Bennington V. 2013. Simulation of heavy lake-effect snowstorms across the Great Lakes Basin by RegCM4: Synoptic climatology and variability. *Monthly Weather Review.* **141**: 1990-2014, doi:10.1175/MWR-D-11-00369.1.

Notaro M, Lorenz D, Hoving C, Schummer M. 2014. Twenty-First-Century Projections of Snowfall and Winter Severity across Central-Eastern North America. *Journal of Climate.* **27**(17): 6526-6550.

Peace RL Jr., Sykes RB Jr. 1966. Mesoscale study of a lake effect snow storm. *Monthly Weather Review.* **94**(8): 495-507.

Pease SR, Lyons WA, Keen CS, Hjelmfelt M. 1988. Mesoscale spiral vortex embedded within a Lake Michigan snow squall band: High resolution satellite observations and numerical model simulations. *Monthly Weather Review.* **116**(6): 1374-1380.

Pradhanang SM, Anandhi A, Mukundan R, Zion MS, Pierson DC, Schneiderman EM, Matonse A, and Frei AP, 2011. Application of SWAT model to assess snowpack development and streamflow in the Cannonsville watershed, New York, USA. *Hydrological Processes.* **25**:3268-3277, doi:10.1002/hyp.8171.

Riggs GA, Hall DK, Román MO. 2016. MODIS Snow Products User Guide for Collection 6, http://modis-snow-ice.gsfc.nasa.gov/?c=userguides.

Riggs GA, Hall DK, Román MO. 2017. Overview of NASA's MODIS and VIIRS Snow Cover Earth System Data Records. *Earth Systems Science Data*, submitted for publication.

Schmidlin TW. 1992. Does lake-effect snow extend to the mountains of West Virginia? *Proc. 49th Annual Eastern Snow Conference,* Oswego, NY, CRREL: 145-148.

Suriano ZJ, Leathers DJ. 2016. Twenty-first century snowfall projections within the eastern Great Lakes region: detecting the presence of a lake-induced snowfall signal in GCMs. *International Journal of Climatology*. **36**: 2200-2209, doi:10.1002/joc.4488.

Suriano ZJ, Leathers DJ. 2017. Synoptic climatology of lake-effect snowfall conditions in the eastern Great Lakes Region. *International Journal of Climatology*. doi:10.1002/joc.5093.

Vermote EF, El Saleous NZ, Justice CO. 2002. Atmospheric correction of MODIS data in the visible to middle infrared: first results. *Remote Sensing of Environment*. **83**(1-2): 97-111.

Villani J, Jurewicz ML, Krekeler J. 2010. The inland extent of lake-effect snow (LES) bands, *Proc. of the 18th GLOMW*. Toronto, Ontario, 22-24 March 2010, http://www.slideserve.com/pearl/the-inland-extent-of-lake-effect-snow-les-bands, last accessed 7 September 2016.

Villani JP, Jurewicz ML Sr., Reinhold K. 2017: Forecasting the inland extent of lake effect snow bands downwind of Lake Ontario. *Journal of Operational Meteorology* **5**(5): 53-70, doi: http://doi.org/10.15191/nwajom.2017.0505.

Wan Z, Zhang Y, Zhang Q, Li Z-L. 2002. Validation of the land-surface temperature products retrieved from Terra Moderate Resolution Imaging Spectroradiometer data. *Remote Sensing of Environ*ment. **83**: 163–180.

Wang J, Bai K, Hu H, Clites A, Colton M, Lofgren B. 2012. Temporal and spatial variability of Great Lakes ice cover, 1973-2010. *Journal of Climate*. **25**: 1318-1329.

Poster Session

NASA Suomi-NPP VIIRS Snow Cover Data Products

GEORGE RIGGS,[1] DOROTHY K. HALL,[2] AND MIGUEL O. ROMÁN[3]

ABSTRACT

A brief description of the NASA Suomi-NPP (S-NPP) Visible Infrared Imaging Radiometer Suite (VIIRS) snow cover data products is presented. The VIIRS snow cover data products are by design similar to the MODerate-resolution Imaging Spectroradiometer (MODIS) snow cover data products to enable development of a consistent long term snow cover extent (SCE) Earth system data record (ESDR).

Keywords: S-NPP, VIIRS, NDSI, snow cover.

INTRODUCTION

The VIIRS snow cover products are produced in sequence beginning with a swath at a nominal pixel spatial resolution of 375 m with nominal swath coverage of 6400 pixels (across track) by 6464 pixels (along track), consisting of 6 minutes of VIIRS scans, and ending with a daily global gridded product. Data products in the Earth Observation Data Information System (EOSDIS) are given an Earth Science Data Type (ESDT) name. The ESDT names and descriptions of the VIIRS snow cover products are given in Table 1. These data products are produced in the Land Science Investigator-led Processing System (LSIPS) described at *https://earthdata.nasa.gov/about/science_investigator-led-processing-systems/sips-snn--land*. The National Snow and Ice Data Center (NSIDC) is the Distributed Active Archive Center (DAAC) distributing and providing user support services for the products (VIIRS Collection 1 doi:10.5067/VIIRS/VNP10.001)

Table 1. Listing of the NASA VIIRS snow-cover products.

Product	ESDT	Description
Snow Cover L2 Swath	VNP10	VIIRS/NPP Snow Cover 6-Min L2 Swath 375 m
Snow Cover L3 Daily Tile	VNP10A1	VIIRS/NPP Snow Cover Daily L3 Global 375 m SIN Grid
Snow Cover L3 CMG*	VNP10C1	VIIRS/NPP Snow Cover Daily L3 Global 0.05Deg CMG

*Climate Modeling Grid

[1] SSAI, NASA/GSFC Terrestrial Information Systems Laboratory, Greenbelt, MD, george.a.riggs@nasa.gov.

[2] Earth System Science Interdisciplinary Center (ESSIC), University of Maryland, College Park, MD, dorothy.k.hall@nasa.gov.

[3] NASA/GSFC Terrestrial Information Systems Laboratory, Greenbelt, MD, miguel.o.roman@nasa.gov.

The ESDT name is used to identify the snow data products. The ESDT also indicates the spatial and temporal processing that has been applied to the data product. The data product levels briefly described are: Level 1B (L1B) is a swath (scene) of VIIRS observation data geolocated to latitude and longitude. A Level 2 (L2) product is a geophysical product that remains in latitude and longitude orientation of L1B. There are Level 2 gridded (L2G) algorithms that map swath products on to the sinusoidal projection for VIIRS land products creating intermediate products used to generate the level 3 products. The L2G intermediate products are not archived. A level 3 (L3) product is a geophysical product that has been temporally and or spatially manipulated, and is in a gridded map projection format and comes as a tile of the global grid. The VIIRS L3 snow products are in the sinusoidal projection for VNP10A1 and in geographic projection for VNP10C1.

VIIRS LEVEL 2 SNOW COVER PRODUCT (VNP10)

The VNP10 product is in netCDF4/HDF5 format and is compliant with netCDF4 Climate and Forecast (CF) Metadata Conventions, Version 1.6, 5 December 2011, for relevant attributes and to support georeferencing of data.

The NSIDC DAAC released the VNP10 product in July 2017. Only the VNP10 product has been released at this time. The products will be released sequentially as coding of the algorithms and data products proceeds in the LSIPS from L2 to L3 products. Production of products is done in a forward processing stream from when the algorithm is put into production and a reprocessing stream to fill in the time back to the start of VIIRS data acquisition, November 2011. Check with NSIDC for current status of released data products [*http://nsidc.org/data/vnp10/versions/1*].

Snow cover is detected using the normalized-difference snow index (NDSI) algorithm which is described and discussed in Riggs et al. (2016a and b). Snow typically has high visible reflectance across 0.3 – 1.0 µm and low shortwave-infrared about 1.6 µm which produce NDSI values in the range of 0.1 – 1.0 for snow cover. For VIIRS the NDSI is calculated as; NDSI = ((band I1 – band I3)/(band I1 + band I3)), with I1 at .64 µm and I3 at 1.6 µm. Since non-snow cover features may also have a positive NDSI value the algorithm includes several data screens to reduce snow cover commission errors associated with non-snow features.

There are two data groups in VNP10 -- the GeolocationData group that has the latitude and longitude data, and the SnowData group that has four snow cover variables. Attributes are attached to the variables to describe the data and file attributes to provide information on provenance and archiving of the product. An example of the VNP10 snow cover variables is presented in Figure 1. The NDSI_Snow_Cover variable (Fig.1A) is the snow map determined from the snow cover detection algorithm. Snow cover is represented by normalized-difference snow index (NDSI) values in the range of 1 – 100. Clouds are masked using the VIIRS cloud mask product (VNP35) cloud confidence flag data. The Basic_QA value (Fig. 1C) provides a general assessment of the algorithm result. The Basic QA value is set to value of other when the solar zenith angle exceeds 70°, the yellow region in Fig. 1C, meaning that lower illumination may reduce the accuracy of the result.

In the snow cover algorithm data screens are applied for the purpose of reducing errors and flagging uncertain results under certain conditions. A discussion of how to interpret and potentially use the Basic_QA and Algorithm_QA_bit _flags is presented in Riggs et al. (2016b and c). There are six Algorithm_QA_bit flags that can be set in the algorithm. The flags can be set for conditions of inland water, low visible reflectance, low NDSI value, combined surface temperature and height data screen, high short wave infrared reflectance and solar zenith angle > 70°. Multiple flags may be set for each pixel. A select combination of flags is shown in Fig. 1D to highlight low visible reflectance conditions and low NDSI values, flagged by bits 1 and 2 respectively, in the swath. Where the high solar zenith flag is on is also shown in Fig. 1D. These algorithm QA flags should be used to evaluate the NDSI snow map and the NDSI data. Discussion and use of the data screens and bit flags is presented in Riggs et al. (2017 and 2016a). To enable users to investigate the NDSI data independently or in combination with the other variables, the NDSI data for all land and inland water, and not masked for cloud (Fig. 1B), are included as a variable. A potential use of the NDSI data and algorithm bit flags is to modify the NDSI snow cover map by replacing pixels that were changed to "not snow" in specific conditions, for example a pixel that is "not snow" and has the low visible reflectance data screen set to on could be changed to "snow" by using the corresponding NDSI data value to replace the "not snow" value.

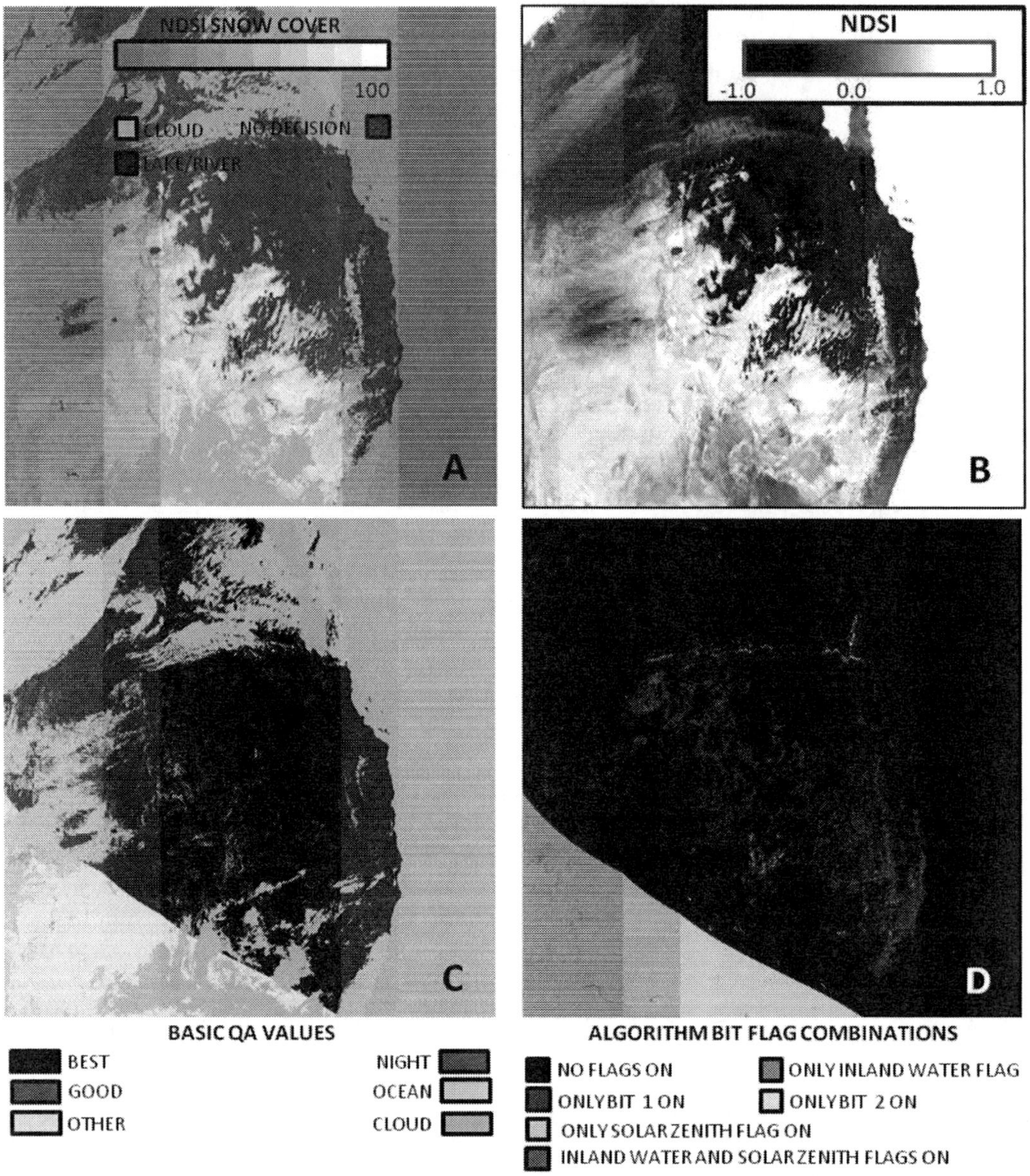

Figure 1. VNP10 acquired 18 December 2016 at 2024 UTC, imaging the western CONUS. Datasets: NDSI_Snow_Cover (A) NDSI (B), Basic QA (C) and a selected combination of the Algorithm_bit_flag QA (D). The image is shown in latitude and longitude orientation thus is not projected. Horizontal lines on sides of the swath are the onboard bowtie trim of overlapping scan lines which is retained in the unprojected image.

VALIDATION

Validation at stage 1 has been achieved for the NASA VIIRS VNP10 snow-cover product. Stage 1 validation is defined as "Product accuracy is assessed from a small (typically < 30) set of locations and time periods by comparison with in-situ or other suitable reference data."

The VIIRS snow-cover product has been compared to SCE and NDSI with both Terra and Aqua MODIS NDSI-derived SCE for various areas in North America. The accuracy of the MODIS snow-cover maps is well established (Hall et al., 2007). VIIRS NDSI snow maps accurately represent the extent of snow cover at regional scales (Riggs et al., 2017).

In a study area that includes the Wind River Range in Wyoming, the agreement between MODIS Collection 6 and VIIRS Collection 1 NDSI-derived SCE maps is 97 percent. The small differences in the location of snow cover mapped by MODIS and VIIRS are due to differences in the sensors, viewing geometry and acquisition times. Results of that study are shown in Fig. 2. In the snow-cover maps (Fig. 2) NDSI values over 0.1 signify snow cover; the deep purple represents the lowest NDSI values and white represents the highest NDSI and thus the most extensive snow cover. The MODIS-VIIRS difference map is shown in the right image of Fig. 2. The small differences in the location of snow cover mapped by MODIS and VIIRS are due to differences in the sensors, viewing geometry and acquisition-times.

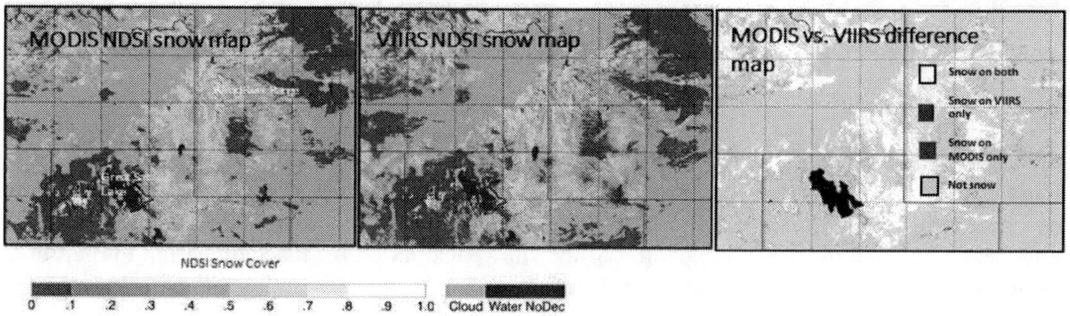

Figure 2. MODIS NDSI snow map focusing on the Wind River Range, Wyoming on 2 December 2016 (left image). VIIRS NDSI snow map covering the same area shown in Figure 1 on 2 December 2016 (center image). A difference map showing agreement between location of snow cover on the MODIS (left image) and VIIRS (center image) snow maps (right image). The light gray shows cloud masking in the difference map (using the MODIS and VIIRS cloud masks combined) and snow-free land.

Status of validation of the products is posted at the NASA S-NPP VIIRS Land website, *https://viirsland.gsfc.nasa.gov/Products/NASA/CryoESDR.html*.

REFERENCES

Hall DK, Riggs GA. 2007. Accuracy assessment of the MODIS snow-cover products, *Hydrological Processes*, **21**(12): 1534-1547. doi:10.1002/hyp.6715.

Riggs GA, Hall DK, Román MO. 2016a. VIIRS Snow Cover Algorithm Theoretical Basis Document (ATBD), NASA VIIRS project document, available at: http://modis-snow-ice.gsfc.nasa.gov/?c=atbd&t=atbd, last access: 8 March 2017.

Riggs GA, Hall DK, Román MO. 2016b. VIIRS snow products user guide for Collection 1 (C1), available at: http://modis-snow-ice.gsfc.nasa.gov/?c=userguides, last access: 8 March 2017.

Riggs GA, Hall DK, Román MO. 2016c. MODIS snow products user guide for Collection 6 (C6), available at: http://modis-snow-ice.gsfc.nasa.gov/?c=userguides, last access: 8 March 2017.

Riggs GA, Hall DK, Román MO. 2017. Overview of NASA's MODIS and Visible Infrared Imaging Radiometer Suite (VIIRS) snow-cover Earth System Data Records, *Earth System Science Data*, **9**: 1–13. doi.org/10.5194/essd-9-1-2017.

A Summary of Snow Research Activities Involving Undergraduate Student Volunteers at Nipissing University

KRYSTOPHER J. CHUTKO,[1] APRIL L. JAMES,[2] AND BRITTANY GERMAIN[2]

EXTENDED ABSTRACT

Keywords: Snow, undergraduate research, isotope hydrology

In January 2014, researchers and students in the Department of Geography at Nipissing University initiated a snow hydrology research project designed to combine high resolution isotopic sampling of snow with extracurricular undergraduate teaching and learning. The goal of the research was to understand how the isotopic composition of the snowpack varied from accumulation to melt in an effort to establish a snowmelt end-member for further study iof catchment hydrology on the Canadian Shield in Ontario, Canada. Physical snow properties were measured and snow/meltwater samplers were collected from bulk cores, snow pits, a snow measurement board, and snowmelt lysimeters. An on-site weather station, including an ultrasonic snow depth sensor, added meteorological data.

The study site was an undeveloped parcel of land on the Nipissing University campus. The site is formerly a century-old farm homestead, and much of the land has been unmanaged for a number of years; natural vegetation has re-established itself over much of the land. Currently, a TransCanada oil pipeline passes through the property, although instrumentation and fieldwork avoided this regularly thawed area. The site is composed of mixed woodland on Boreal Shield terrain, typical of the surrounding region.

Specific siting of fieldwork differed between the 2014 and the 2015-16 fieldwork seasons (Fig. 1). In 2014, an Onset HOBO meteorological station was paired with a Campbell Scientific SR50 sonic ranging sensor which collected continuous snow depth measurements at 10 minute intervals throughout the snow-on-the-ground season. A snow measurement board was co-located at the meteorological station and sampled at regular intervals. After depth measurements and snow samples were taken, the board was cleaned and placed on top of the snowpack to record "event" snow. A 100m snow course, oriented NNE-SSW was established nearby, as was an undisturbed area for repeated snow pits. Bulk and stratified snow samples were obtained along the snow course at 10 m intervals at least once per week from January through April. Additional sampling occurred after snowfall events or during the melt period. Snow pits were dug approximately 2 m apart, with a sampling interval concurrent with the snow board and snow course.

[1] Department of Geography and Planning, University of Saskatchewan, 117 Science Place, Saskatoon, SK S7N 5C8 krys.chutko@usask.ca

[2] Department of Geography, Nipissing University, 100 College Dr., North Bay, ON P1B 8L7 http://iwrc.nipissingu.ca

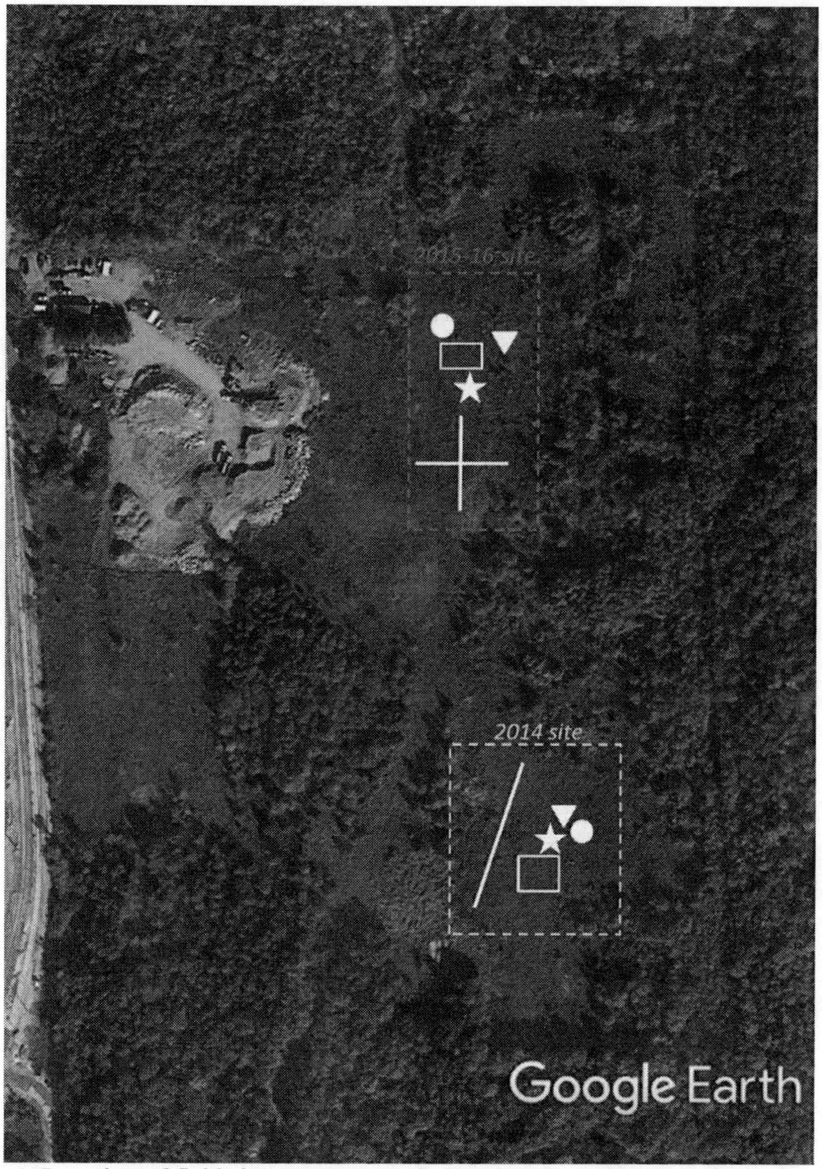

Figure 1. Location of field site components during the 2014-16 period. Symbols: star = meteorological and other sensor stations; circle = snow board; triangle = snowmelt lysimeters; open box = snow pit area; lines = snow courses.

For the 2015-16 winter seasons, the snow board, snow course, and snow pit area were moved approximately 300 m north, while the meteorological station remained at the original site. A new snow measurement station was established at the new site, which included the SR50 ultrasonic depth sensor and soil moisture probes. The snow course was oriented in the shape of a cross, with N-S and E-W branches.

Snowmelt lysimeters were constructed at each site for the three years of study. The structure of the lysimeters changed from year-to-year depending on observations of snowmelt patterns and meltwater catch efficacy. The original structure was a 2m x 2m x 10cm plywood box; this structure was too large, resulting in excessive meltwater and rapid overfilling of sampling containers. In 2016, a new pair of snowmelt lysimeters using 1 m diameter plastic pans were used which improved both the representativeness of the sample and ability to catch a reasonable amount of meltwater.

Snowmelt lysimeter construction and use was hampered considerably by the low relief of the field site.

Figure 2. A: a view of the 2014 site looking NNE. The snow course extends along the left edge of the photograph. The snow pit area is to the right of the meteorological station in the centre background. B: the meteorological station at the 2014 site. The SR50 snow depth sensor is position on the white frame behind the station. C: the sensor station at the 2015-16 site. The Onset HOBO meteorological station continue to run at the original location in 2015-16. In addition to measuring snow depth and air temperature (a key component of the ultrasonic measurement), two soil pits containing soil moisture sensors were included. D: snowmelt lysimeters installed at the 2015-16 site. Due to low relief at the site, a trench was dug to contain the meltwater container. Water samples were collected during melt events (up to twice daily), volumes measured, and subsampled for isotopic analysis.

An invaluable component of the research project included a group of volunteer undergraduate students who participated in the semi-weekly sampling program, running concurrent with regular course work. On each sampling day, a group of 2 – 4 students was organized via Facebook. At the sampling site, student would work together collecting bulk density measurements and snow samples along the snow course, and qualitative (e.g., presence of ice layers, grain morphology) and quantitative (e.g., layer thickness and density) measures of snowpack stratigraphy from snow pits. Impromptu lessons on sampling strategies, snow hydrology, and stable water isotope chemistry, at the request of the students, often accompanied the sampling. The efforts by the students have contributed to a multi-year documentation of snowpack stable isotope evolution and overall watershed-scale isotopic studies in the region.

Figure 3. Student involvement in all aspects of snow sampling was a critical component of this project. In 2014, winter temperatures at the site reached -40 °C, revealing to students some of the challenges faced when conducting winter research.

While measuring in situ snowpack properties were an important component of this research, understanding the isotopic evolution of the snowpack through the snow-on-the-ground season was the primary interest (Fig. 4). During measurement excursions, several samples of snow – bulk cores from the snow course, 10 cm depth intervals from the snow pits, and event snow from the snow board – were collected, slowly melted at 4 °C to minimize fractionation processes, and subsampled for analysis. All snow samples were analyzed at the Integrative Watershed Research Centre at Nipissing University using a Picarro L2120-I cavity ring-down spectrometer.

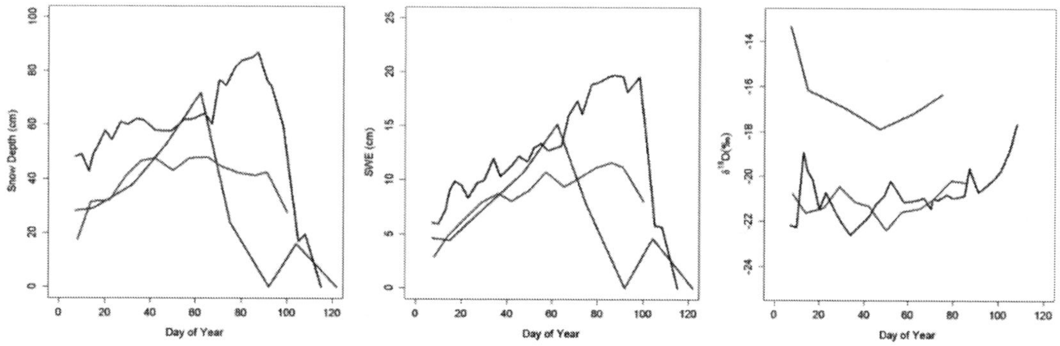

Figure 4: Left-to-right: snow depth, snow water equivalent (SWE), and $\delta^{18}O$ results for 2014-2016 snow measurement years. These data were collected primarily by volunteer undergraduate students, many of whom participated in more than one year. In 2016, a relatively warm winter resulted in substantially different snow properties, particularly related to oxygen and hydrogen isotopes. These data are now used in other studies to understand the role that snowmelt waters hold in seasonally variable isotopic composition of streamflow in the southern Boreal region of Ontario.

ACKNOWLEDGEMENTS

This work was financially supported by the Canada Research Chair program, the Canada Foundation of Innovation, and Nipissing University. Special thanks to hydrology technicians David McCorkell, Melodie Tessier, and Megan Prescott., and to the students: Duncan McLeod, Jackie Saunders, Christian Peart, Jasmine Taulu, Rebecca Garlick, Markus Fleming, Sam Armstrong, Celine Larocque, Kiera Vere, Emily Selander, Leah Tomblin, Julie Culligan, Mark Postpischl, Carley Zimba, and Dana Lacivita.

SNO-FOO 2017 – Michael Durand

The 2017 SNO-FOO award is presented to Dr. Michael Durand, Associate Professor in the School of Earth Science at The Ohio State University. Mike wins the award while simultaneously serving as President of the Eastern Snow Conference, a not-so-enviable feat!

Mike Durand is an exceptional collaborator. Enthusiastic to the core, and always willing to get stuck into an interesting problem. As a consequence, he is usually phenomenally busy. This was even more so the case in the months leading up to him obtaining tenure at The Ohio State University. We were very fortunate to spend some time with him at the Snow Grain Size Workshop in Grenoble, France, April 2013. This cozy meeting was focused on (and resulted in) making significant progress in measurement of snow microstructure, and its representation in radiative transfer modelling. In other words, bang on Mike's research interests. A critical meeting

Not wanting to abrogate his lecture responsibilities in favour of leisure time in the French Alps, he elected to spend the bare minimum of time away from his desk and planned to miss the first day of the meeting. We were somewhat surprised not to see Mike on day two though, Alas, a flight cancellation and a missed London to Grenoble train connection (by 5 minutes) meant that he ended up avoiding all of the science presentations. He did, however, make it in time for his transatlantic to consist of: a) a fun field trip to La Grave, and b) three hours of science discussion. This award is make in recognition of his dedication to his students, optimism in the face of adversity, and sheer persistence.

Presented by Melody Sandells